Statistical Mecha
A Concise Introduction fo

Statistical mechanics is the theoretical apparatus used to study the properties of macroscopic systems – systems made up of many atoms or molecules – and relates those properties to the systems' microscopic constitution. This book is an introduction to statistical mechanics, intended to be used either by advanced undergraduates or by beginning graduate students.

The first chapter deals with statistical thermodynamics and aims to derive quickly the most commonly used formulas in the subject. The bulk of the book then illustrates the traditional applications of these formulas, such as the ideal gas, chemical equilibrium constants from partition functions, the ideal harmonic solid, and the statistical mechanical basis of the third law of thermodynamics. The last few chapters deal with less traditional material such as the non-ideal gas and the quantum ideal gases.

BENJAMIN WIDOM is Goldwin Smith Professor of Chemistry at Cornell University. He received his Ph.D. in physical chemistry from that university (where he studied with S.H. Bauer) in 1953, and was a postdoctoral associate with O.K. Rice at the University of North Carolina, before joining the Cornell chemistry faculty in 1954.

His research specialty is statistical mechanics and thermodynamics, particularly as applied to problems of phase equilibria, critical phenomena, and interfacial structure and thermodynamics. He is co-author with Professor Sir John Rowlinson, of Oxford University, of the research monograph *Molecular Theory of Capillarity* (1982).

He has held numerous prestigious visitorships, including ones at Amsterdam (van der Waals Professor), Oxford (IBM Visiting Professor of Theoretical Chemistry), Leiden (Lorentz Professor), and Utrecht (Kramers/Debye Professor). He has had many awards in recognition of his research in statistical mechanics, including the Boltzmann Medal of the IUPAP Commission on Statistical Physics and the Onsager Medal of the University of Trondheim. He has honorary degrees from the University of Chicago and the University of Utrecht, and has been elected to membership or fellowship of several scholarly academies including the U.S. National Academy of Sciences and the American Philosophical Society.

Statistical Mechanics
A Concise Introduction for Chemists

B. Widom
Cornell University

PUBLISHED BY THE PRESS SYNDICATE OF THE UNIVERSITY OF CAMBRIDGE
The Pitt Building, Trumpington Street, Cambridge, United Kingdom

CAMBRIDGE UNIVERSITY PRESS
The Edinburgh Building, Cambridge CB2 2RU, UK
40 West 20th Street, New York, NY 10011-4211, USA
477 Williamstown Road, Port Melbourne, VIC 3207, Australia
Ruiz de Alarcón 13, 28014 Madrid, Spain
Dock House, The Waterfront, Cape Town 8001, South Africa

http://www.cambridge.org

© Benjamin Widom 2002

This book is in copyright. Subject to statutory exception
and to the provisions of relevant collective licensing agreements,
no reproduction of any part may take place without
the written permission of Cambridge University Press.

First published 2002

Printed in the United Kingdom at the University Press, Cambridge

Typeface Times 10/13 pt.   *System* LATEX 2_ε [TB]

A catalogue record for this book is available from the British Library

Library of Congress Cataloguing in Publication data
Widom, B.
Statistical mechanics / B. Widom.
p.   cm.
Includes bibliographical references and index.
ISBN 0 521 81119 8 – ISBN 0 521 00966 9 (pb.)
1. Statistical mechanics.   I. Title.
QC174.8 .W53   2002
530.13–dc21   2001043687

ISBN 0 521 81119 8 hardback
ISBN 0 521 00966 9 paperback

Contents

Preface		*page* vii
1	The Boltzmann distribution law and statistical thermodynamics	1
	1.1 Nature and aims of statistical mechanics	1
	1.2 The Boltzmann distribution law	2
	1.3 The partition function and statistical thermodynamics	6
2	The ideal gas	16
	2.1 Boltzmann statistics	16
	2.2 Translational partition function	21
	2.3 Vibrational partition function	27
	2.4 Rotational partition function; *ortho*- and *para*-hydrogen	31
	2.5 The "law" of the equipartition of energies	41
	2.6 Partition function with excited electronic states	43
3	Chemical equilibrium in ideal-gas mixtures	47
	3.1 Thermodynamic preliminaries; the equilibrium constant	47
	3.2 Equilibrium constants from partition functions	49
4	Ideal harmonic solid and black-body radiation	55
	4.1 Ideal harmonic crystal	55
	4.2 Rayleigh–Jeans law	57
	4.3 Debye theory of the heat capacity of solids	62
	4.4 Black-body radiation	66
5	The third law	69
	5.1 Nernst heat theorem in thermodynamics	69
	5.2 Third law in statistical mechanics	71
	5.3 Comparison with experiment	75
6	The non-ideal gas	81
	6.1 Virial coefficients	81
	6.2 Intermolecular forces	83
	6.3 Second virial coefficient from statistical mechanics	87
	6.4 Long-range forces	98

7	The liquid state	101
	7.1 Structure of liquids	101
	7.2 Equation of state of a liquid	106
	7.3 Computer simulation: molecular dynamics	114
	7.4 Computer simulation: Monte Carlo	127
8	Quantum ideal gases	133
	8.1 Bose–Einstein and Fermi–Dirac statistics versus Boltzmann statistics	133
	8.2 The grand-canonical partition function	138
	8.3 Grand partition function of the quantum ideal gases	143
	8.4 The ideal gas in Fermi–Dirac statistics	155
	8.5 The ideal gas in Bose–Einstein statistics	162
Index		169

Preface

This is an introduction to statistical mechanics, intended to be used either in an undergraduate physical chemistry course or by beginning graduate students with little undergraduate background in the subject. It assumes familiarity with thermodynamics, chemical kinetics and the kinetic theory of gases, and quantum mechanics and spectroscopy, at the level at which these subjects are normally treated in undergraduate physical chemistry. Ideas, principles, and formulas from them are appealed to frequently in the present work.

If statistical mechanics constituted about 10% of a physical chemistry course it would be covered in 8 to 12 lectures, depending on whether the course as a whole were taught in two semesters or three. There is enough material in these chapters for 12 lectures (or more). The instructor who has only 8 available will have to be selective. The most technical parts, and so the likeliest candidates for omission or contraction, are the treatment of *ortho-* and *para*-hydrogen, which is part of §2.4 of Chapter 2, that of molecular dynamics and Monte Carlo computer simulations in Chapter 7 (§§7.3 and 7.4), and that of the quantum ideal gases in Chapter 8, which includes a discussion of the grand partition function (§8.2).

Because only a relatively short time may be devoted to the subject it is important to arrive quickly at the usable formulas and important applications while still keeping the level consistent with that of an undergraduate physical chemistry course. The strategy adopted here is to start with the Boltzmann distribution law, making it plausible by appeal to two of its special cases (the Maxwell velocity distribution, assumed to be known from an earlier treatment of elementary kinetic theory, and the barometric distribution, which is derived, or re-derived, here), and by observing that its exponential form is required by the composition of probabilities for independent systems. The distribution law is stated with discrete-state, quantum mechanical energy-level notation at an early stage so as not to require generalizing the theory later from classical to quantum mechanical language. In a full-scale, graduate-level statistical mechanics course, where one has the leisure to devote more time to the foundations and where the students have more background in physics and mathematics, one can make a good case for starting with a classical phase-space formulation and later generalizing it to the quantum mechanical version. One would then typically start with the microcanonical rather than the canonical distribution.

Once the Boltzmann distribution law is stated the partition function appears naturally as a normalization denominator. The machinery of statistical thermodynamics – the connection of the free energy to the partition function – then follows from comparing the mean energy implied by the distribution law with that obtained from the free energy by the Gibbs–Helmholtz equation of thermodynamics. (Among earlier treatments of the subject, that in Guggenheim's wonderful little book *Boltzmann's Distribution Law* is probably that to which this one is closest in spirit.) This is followed by a qualitative discussion of energy fluctuations in a system of fixed temperature, and then the connection is made to a microcanonical system and the famous $S = k \ln W$.

What has just been outlined is the content of the first chapter, in which the machinery is set up. The next four chapters are devoted to the traditional applications: the statistical thermodynamics of the ideal gas of molecules with internal structure (Chapter 2), chemical equilibrium constants from partition functions (Chapter 3), the ideal harmonic solid (Chapter 4), and the statistical mechanical basis of the third law of thermodynamics (Chapter 5).

Much of the material in the remaining three chapters is less traditional in an undergraduate physical chemistry course but would be suitable even there and certainly at the beginning graduate level. The non-ideal gas (Chapter 6) provides the first glimpse of the problem of non-separable degrees of freedom, which appears in its undiluted and most challenging form in the liquid state (Chapter 7). The concluding chapter, on the quantum ideal gases, includes the degenerate electron gas as a model for electrons in metals, a subject that most instructors would wish to say something about even if they should choose to omit much of the detail in that section and much of the rest of the chapter.

Only equilibrium statistical mechanics is presented here. There are many topics in non-equilibrium statistical mechanics (the Nernst–Einstein relation between diffusion coefficient and mobility being an example) that are of interest and importance in physical chemistry, but given only 8–12 lectures some heartbreaking choices have to be made, as every instructor knows.

I wish to express my appreciation to Professors Roger Loring (Cornell), Igal Szleifer (Purdue), and Devarajan Thirumalai (Maryland) for their advice and encouragement, and to Ms Kelly Case for her expert assistance with the preparation of the manuscript.

B. Widom
Cornell University

1
The Boltzmann distribution law and statistical thermodynamics

1.1 Nature and aims of statistical mechanics

Statistical mechanics is the theoretical apparatus with which one studies the properties of macroscopic systems – systems made up of many atoms or molecules – and relates those properties to the system's microscopic constitution. One branch of the subject, called statistical thermodynamics, is devoted to calculating the thermodynamic functions of a system of given composition when the forces of interaction within and between the system's constituent molecules are given or are presumed known. This first chapter is directed toward obtaining the most commonly used formulas of statistical thermodynamics and much of the remainder of the book illustrates their application.

Because the systems to which the theory is applied consist of large numbers of molecules, and are thus systems of a large number of mechanical degrees of freedom, we are not interested in all the details of their underlying microscopic dynamics (and could hardly hope to know them even if we were interested). Instead, it is the systems' macroscopic properties – among which are the thermodynamic functions – that we wish to understand or to calculate, and these are gross averages over the detailed dynamical states. That is the reason for the word "statistical" in the name of our subject.

A prominent feature in the landscape of statistical mechanics is the Boltzmann distribution law, which tells us with what frequency the individual microscopic states of a system of given temperature occur. An informal statement of that law is given in the next section, where it is seen to be an obvious generalization of two other well known distribution laws: the Maxwell velocity distribution and the "barometric" distribution. We also remark there that the exponential form of the Boltzmann distribution law is consistent with – indeed, is required by – the rule that the probability of occurrence of independent events is the product of the separate probabilities.

1 Statistical thermodynamics

In §1.3 we shall find that the normalization constant that occurs in the Boltzmann distribution is related to the system's free energy. That is the key to statistical thermodynamics. Together with a related but simpler observation about the connection between thermodynamic and mechanical energy, it amounts to having found the microscopic interpretation of the first and second laws of thermodynamics.

1.2 The Boltzmann distribution law

The Boltzmann distribution law says that if the energy associated with some state or condition of a system is ε then the frequency with which that state or condition occurs, or the probability of its occurrence, is proportional to

$$e^{-\varepsilon/kT}, \tag{1.1}$$

where T is the system's absolute temperature and where k is the Boltzmann constant, which the reader will already have encountered in the kinetic theory of gases:

$$k = 1.38 \times 10^{-23} \text{ J/K} = 1.38 \times 10^{-16} \text{ erg/K}. \tag{1.2}$$

Many of the most familiar laws of physical chemistry are obvious special cases of the Boltzmann distribution law. An example is the Maxwell velocity distribution. Let v be one of the components of the velocity (v_x or v_y or v_z) of a molecule in a fluid (ideal gas, imperfect gas, or liquid – it does not matter), let m be the mass of the molecule, and let $f(v)dv$ be the probability that v will be found in the infinitesimal range v to $v + dv$. This $f(v)$ is one of the velocity distribution functions that play a prominent part in the kinetic theory of gases. A graph of $f(v)$ is shown in Fig. 1.1. Roughly speaking, it gives the frequency of occurrence of the value v for that chosen velocity component. More precisely, the probability $f(v)dv$ (which is the area under the $f(v)$ curve

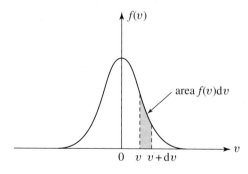

Fig. 1.1

1.2 The Boltzmann distribution law

between v and $v + dv$; Fig. 1.1) is the fraction of the time, averaged over long times, that any one molecule has for that one of its velocity components a value in the range v to $v + dv$. (The velocities of the individual molecules continually change with time because the molecules continually collide and otherwise interact with each other.) Alternatively but equivalently, at any one instant the infinitesimal fraction $f(v)dv$ of all the molecules are molecules that have for that component of their velocity a value in the range v to $v + dv$. The velocity distribution function $f(v)$ is

$$f(v) = \sqrt{\frac{m}{2\pi kT}}\, e^{-mv^2/2kT}. \tag{1.3}$$

The energy associated with that velocity component's having the value v (kinetic energy in this instance) is $\varepsilon = \frac{1}{2}mv^2$, so the Maxwell velocity distribution (1.3) is obviously a special case of the Boltzmann distribution (1.1).

Another special case of the Boltzmann distribution (1.1) is the "barometric" distribution, giving the number density $\rho(h)$ (number of molecules per unit volume) of an ideal gas of uniform temperature T as a function of height h in the field of the earth's gravity. (This could be the earth's atmosphere, say, with the temperature assumed not to vary too much with height – although that is a questionable assumption for the atmosphere.) A column of such a gas of arbitrary cross-sectional area A is depicted in Fig. 1.2. The volume of that part (shown shaded in the figure) that is between h and $h + dh$ is Adh and its mass is then $m\rho(h)\,Adh$ with m again the mass of a molecule. With the gravitational acceleration g acting downward, that infinitesimal element of the gas, because of its weight, exerts a force $mg\,\rho(h)\,Adh$ on the column of gas below it. The excess pressure (force per unit area) at the height h over that at the height $h + dh$ is then

$$p_h - p_{h+dh} = -dp = mg\,\rho(h)dh. \tag{1.4}$$

This would be true no matter what fluid was in the column; but for an ideal gas the pressure p and number density ρ are related by $p = \rho kT$. [This follows

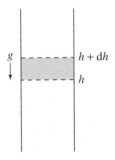

Fig. 1.2

from the ideal-gas law $pV = nRT$, with n the number of moles and R the gas constant, together with $n = N/N_0$, where N is the number of molecules and N_0 is Avogadro's number, with $k = R/N_0$ (as identified in the kinetic theory of the ideal gas), and with $\rho = N/V$.] Therefore (1.4) becomes a differential equation for $\rho(h)$,

$$d\rho(h)/dh = -(mg/kT)\rho(h). \tag{1.5}$$

The solution, as we may readily verify by differentiation with respect to h, is

$$\rho(h) = \rho(h_0)e^{-mg(h-h_0)/kT} \tag{1.6}$$

where h_0 is an arbitrary fixed reference height. If the gas is a mixture of different species with differing molecular masses m, each has its own distribution (1.6) with its own m.

This is the barometric distribution. Since the probability of finding any specified molecule at the height h is proportional to the number density there, (1.6) is equally well the probability distribution as a function of the height h. Then (1.6) says that the probability of finding a specified molecule at h varies with h as $\exp(-mgh/kT)$. But we recognize mgh as the energy ε (potential energy in this instance) associated with the molecule's being in that state – i.e., at the height h in the earth's gravity. Thus, (1.6) is clearly another special case of the Boltzmann distribution law (1.1).

Exercise (1.1). Spherical particles of diameter 0.5 μm and density 1.10 g/cm^3 are suspended in water (density 1.00 g/cm^3) at 20 °C. (Particles of such size are called colloidal and a suspension of such particles is a colloid or a colloidal suspension.) Find the effective mass m of the particles, corrected for buoyancy, and then calculate the vertical distance over which the number density of suspended particles decreases by the factor $1/e$. (Historically, in experiments by J. Perrin, who measured the distribution of such particles with height, this was one of the methods by which Boltzmann's constant k was determined – or, equivalently, Avogadro's number N_0, since $N_0 = R/k$ and the gas constant R is known from simple measurements on gases.)

Solution. The volume of each particle is $(4\pi/3)(0.5/2)^3 \, 10^{-18}$ m$^3 = 0.065 \times 10^{-12}$ cm^3, so the effective mass is $m = (0.065 \times 10^{-12})(1.10 - 1.00)$ g $= 6.5 \times 10^{-15}$ g. From the barometric distribution (1.6), the vertical distance over which the number density decreases by the factor e is kT/mg, which, with $k = 1.38 \times 10^{-16}$ erg/K, $T = 293$ K, $m = 6.5 \times 10^{-15}$ g, and acceleration due to gravity, $g = 981$ cm s^{-2}, is 6.3×10^{-3} cm.

1.2 The Boltzmann distribution law

The exponential dependence of the probability on the energy in the distribution law (1.1) is a reflection of the product law for the composition of probabilities of independent events. Suppose ε_1 is the energy associated with some state or condition of the system and ε_2 is that associated with some other condition, and that the occurrences of these two states are independent. For example, we might ask for the chance that one molecule in a fluid has its x-component of velocity in the range v_1 to $v_1 + dv_1$, for which the associated energy is $\varepsilon_1 = \frac{1}{2}mv_1^2$, while another molecule has its x-component of velocity in the range v_2 to $v_2 + dv_2$, for which $\varepsilon_2 = \frac{1}{2}mv_2^2$; or v_1 and v_2 could be the x- and y-components of velocity of the same molecule, these, too, being independent of each other. Then, with the simultaneous occurrence of the two events viewed as a single event, the energy associated with it is $\varepsilon_1 + \varepsilon_2$, while the probability of its occurrence must be the product of the separate probabilities. The probability must therefore be an exponential function of the energy ε, because the exponential is the unique function $F(\varepsilon)$ with the property $F(\varepsilon_1 + \varepsilon_2) = F(\varepsilon_1)F(\varepsilon_2)$:

$$e^{-(\varepsilon_1+\varepsilon_2)/kT} = e^{-\varepsilon_1/kT}\, e^{-\varepsilon_2/kT}. \tag{1.7}$$

That the parameter determining how rapidly the exponential decreases with increasing energy is the absolute temperature is a law of nature; we could not have guessed that by mathematical reasoning alone.

For the purposes of developing statistical thermodynamics in the next section we shall here apply the distribution law to tell us the frequency of occurrence of the states i, of energy E_i, of a whole macroscopic system. For generality we may suppose these to be the quantum states (although classical mechanics is often an adequate approximation). We should understand, however, that there may be tremendous complexity hidden in the simple symbol i. We may think of it as a composite of, and symbolic for, some enormous number of quantum numbers, as many as there are mechanical degrees of freedom in the whole system, a number typically several times the number of molecules and thus perhaps of the order of 10^{23} or 10^{24}.

For such a macroscopic system in equilibrium at the temperature T, the probability P_i of finding it in the particular state i is, according to the Boltzmann distribution law (1.1),

$$P_i = \frac{e^{-E_i/kT}}{\sum_i e^{-E_i/kT}}. \tag{1.8}$$

The denominator is the sum of $\exp(-E_i/kT)$ over all states i (and so does not depend on i, which is there just a dummy summation index), and is what guarantees that P_i is properly normalized:

$$\sum_i P_i = 1. \tag{1.9}$$

That normalization denominator is called the *partition function* of the system, and is the key to statistical thermodynamics.

1.3 The partition function and statistical thermodynamics

What we identify and measure as the thermodynamic energy U of a macroscopic system is the same as its total mechanical energy E: the sum total of all the kinetic and potential energies of all the molecules that make up the system. Is that obvious? If it now seems obvious it is only because we have given the same name, energy, to both the thermodynamic and the mechanical quantities, but historically they came to be called by the same name only after much experimentation and speculation led to the realization that they are the same thing.

Two key observations led to our present understanding. The energy E of an isolated mechanical system is a constant of the motion; although the coordinates and velocities of its constituent parts may change with time, that function of them that is the energy has a fixed value, E. That is at the mechanical level. At the thermodynamic level, as one aspect of the first law of thermodynamics, it was recognized that if a system is thermally and mechanically isolated from its surroundings – thermally isolated so that no heat is exchanged ($q = 0$) and mechanically isolated so that no work is done ($w = 0$) – then the function U of its thermodynamic state does not change. That is one fundamental property that the mechanical E and the thermodynamic U have in common. The second is that if the mechanical system is not isolated, its energy E is not a constant of the motion, but can change, and does so by an amount equal to the work done on the system: $\Delta E = w$. Likewise, in thermodynamics, if a system remains thermally insulated ($q = 0$), but is mechanically coupled to its environment, which does work w on it, then its energy U changes by an amount equal to that work: $\Delta U = w$. This coincidence of two such fundamental properties is what led to the hypothesis that the thermodynamic function U in the first law of thermodynamics is just the mechanical energy E of a system of some huge number of degrees of freedom: the total of the kinetic and potential energies of the molecules.

If our system is not isolated but is in a thermostat that fixes its temperature T and with which it can exchange energy, then the energy E is not strictly constant, but can fluctuate. Such energy fluctuations in a system of fixed temperature, while often interesting and sometimes important, are of no thermodynamic consequence: the fluctuations in the energy are minute compared with the total

1.3 The partition function and statistical thermodynamics

and are indiscernible at a macroscopic level. Therefore the thermodynamic energy U in a system of fixed temperature T may be identified with the mean mechanical energy $\bar{E}$ about which the system's mechanical energy fluctuates.

That mean energy $\bar{E}$ of a system of given T is now readily calculable from the Boltzmann distribution (1.8):

$$\bar{E} = \frac{\sum_i E_i e^{-E_i/kT}}{\sum_i e^{-E_i/kT}}, \tag{1.10}$$

and this is what we may now identify as the thermodynamic energy U at the given T. To know the system's energy levels E_i we must know its volume V and also its chemical composition, i.e., the numbers of molecules $N_1, N_2, \ldots$ of each chemical species $1, 2, \ldots$ present in the system, for only then is the mechanical system defined. The energy levels E_i are therefore themselves functions of $V, N_1, N_2, \ldots$, and the $\bar{E}(=U)$ obtained from (1.10) is then a function of these variables and of the temperature T. From the identity $d \ln x/dx = 1/x$ and the chain rule for differentiation, we then see that (1.10) implies

$$U(T, V, N_1, N_2, \ldots) = \bar{E} = -\left(\frac{\partial}{\partial \frac{1}{kT}} \ln \sum_i e^{-E_i/kT}\right)_{V, N_1, N_2, \ldots}. \tag{1.11}$$

The argument of the logarithm in (1.11) is just the normalization denominator in the probability distribution P_i in (1.8). It is called the *partition function*, as remarked at the end of §1.2. It is a function of temperature, volume, and composition. We shall symbolize it by Z, so

$$Z(T, V, N_1, N_2, \ldots) = \sum_i e^{-E_i/kT}. \tag{1.12}$$

Equation (1.11) is then

$$U = -k[\partial \ln Z/\partial(1/T)]_{V, N_1, N_2, \ldots}. \tag{1.13}$$

Now compare this with the Gibbs–Helmholtz equation of thermodynamics,

$$U = [\partial(A/T)/\partial(1/T)]_{V, N_1, N_2, \ldots} \tag{1.14}$$

with A the Helmholtz free energy. We conclude that there is an intimate connection between the free energy A and the partition function Z,

$$A = -kT \ln Z + T\phi(V, N_1, N_2, \ldots) \tag{1.15}$$

where ϕ is some as yet unknown function of just those variables $V, N_1, N_2, \ldots$ that are held fixed in the differentiations in (1.13) and (1.14). Because it is independent of T, this ϕ does not contribute to those derivatives and thus, so far, could be any function of volume and composition.

In the next chapter, §2.2, and in Chapter 3, §3.2, we shall see from what (1.15) implies for an ideal gas that ϕ is in fact independent of V and is an arbitrary linear function of $N_1, N_2, \ldots$, associated with an arbitrary choice for the zero of entropy. (Since $A = U - TS$ with S the entropy, such an arbitrary additive term in the entropy becomes an arbitrary additive multiple of the absolute temperature T in the free energy, as in (1.15).) We shall then follow the universally accepted convention of taking that arbitrary linear function of $N_1, N_2, \ldots$ to be 0. Thus,

$$A = -kT \ln Z. \tag{1.16}$$

In the meantime, since the energy scale also has an arbitrary zero, all the energy levels E_i in the expression (1.12) for Z may be shifted by a common arbitrary amount η, say. There is then an arbitrary factor of the form $\exp(-\eta/kT)$ in Z, which, via (1.13), manifests itself as an arbitrary constant η in U, as expected. The same η also appears as an arbitrary additive constant (in addition to the arbitrary multiple of T) in the free energy A in (1.15) (now associated with the U in $A = U - TS$).

Calculating the partition function Z is the central problem of statistical thermodynamics, and much of the remainder of this book is devoted to calculating it for specific systems. Once the system's partition function has been calculated its Helmholtz free energy A follows from (1.16). That free energy is thus obtained as a function of the temperature, volume, and composition. As a function of just those variables, A is a *thermodynamic potential*; i.e., all the other thermodynamic functions of the system are obtainable from $A(T, V, N_1, N_2, \ldots)$ by differentiations alone, no integrations being required. For example, we have the thermodynamic identities

$$S = -(\partial A/\partial T)_{V, N_1, N_2, \ldots} \tag{1.17}$$

$$U = A + TS \tag{1.18}$$

$$p = -(\partial A/\partial V)_{T, N_1, N_2, \ldots} \tag{1.19}$$

$$\mu_1 = (\partial A/\partial N_1)_{T, V, N_2, N_3, \ldots}, \quad \text{etc.} \tag{1.20}$$

in addition to the Gibbs–Helmholtz equation (1.14), yielding the entropy S, energy U, pressure p, and chemical potentials $\mu_1, \mu_2, \ldots$. (These are molecular rather than molar chemical potentials; they differ from the ones usually introduced in thermodynamics by a factor of Avogadro's number. The molecular chemical potential is the one more frequently used in statistical mechanics, where chemical composition is usually given by numbers of molecules $N_1, N_2, \ldots$ rather than numbers of moles $n_1, n_2, \ldots$.)

1.3 The partition function and statistical thermodynamics

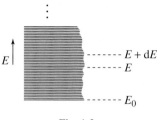

Fig. 1.3

For a macroscopic system, which consists of many, more or less strongly interacting particles, the spacing of the energy levels E_i is usually very much less than the typical thermal energy kT. On any reasonable scale the levels would appear to be almost a continuum. If for no other reason, that would be true because one component of each E_i is the total translational energy of all the molecules, which is then the energy of a large number of particles in a box of macroscopic volume V. But the spacing of particle-in-a-box energy levels decreases with increasing size of the box, as one learns in quantum mechanics, and so is very small when V is of macroscopic size. A consequence of this close spacing of the energy levels E_i is that it is often more convenient and more realistic to treat those levels as though they formed a continuum, and to describe the distribution of the levels as a *density of states*, $W(E)$, such that $W(E)\,dE$ is the number of states i with energies E_i in the infinitesimal range E to $E + dE$. This is illustrated in Fig. 1.3, which shows, schematically, a near continuum of energy levels starting from the ground state of energy E_0. The density of these levels at the energy E, that is, the number of states per unit energy at that E, is $W(E)$.

Since all the states in the infinitesimal energy range E to $E + dE$ have essentially the same energy E, that part of the summation over states i in (1.12) that is over the states with energies in that range contributes to the partition function Z just the common $\exp(-E/kT)$ times the number, $W(E)\,dE$, of those states; and the full sum over i is then the sum (integral) of all these infinitesimal contributions $\exp(-E/kT)W(E)\,dE$. Thus,

$$Z = \int_{E_0}^{\infty} e^{-E/kT} W(E)\,dE. \tag{1.21}$$

The density of states, $W(E)$, depends also on the volume and composition of the system, so, expressed more fully, it is a function $W(E, V, N_1, N_2, \ldots)$. Equation (1.21) then expresses $Z(T, V, N_1, N_2, \ldots)$ as an integral transform (a so-called Laplace transform) of the density of states: multiplying W by $\exp(-E/kT)$ and integrating over all E transforms $W(E)$ into $Z(T)$.

From this same point of view the Boltzmann distribution law, too, may be written in terms of a continuous probability distribution $Q(E)$ rather than in terms of the discrete P_i. The probability $Q(E)\,dE$ that the system will be found in any of the states i with energies E_i in the range E to $E + dE$ is found by summing the P_i of (1.8) over just those states. Again, $\exp(-E_i/kT)$ has the nearly constant value $\exp(-E/kT)$ for each term of the sum and there are $W(E)\,dE$ such terms, so from (1.8) and the definition of Z in (1.12), we have $Q(E)\,dE = Z^{-1}\exp(-E/kT)W(E)\,dE$, or

$$Q(E) = Z^{-1}W(E)e^{-E/kT}. \tag{1.22}$$

This is the form taken by the Boltzmann distribution law when it is recognized that the energies E_i of the states i of a macroscopic system are virtually a continuum with some density $W(E)$.

We have already remarked that when the temperature of a system is prescribed its energy E fluctuates about its mean energy $\bar{E}$ but that the departures of E from $\bar{E}$ are not great enough to be discernible at a macroscopic level, so that the thermodynamic energy U may be identified with $\bar{E}$. Since the probability of finding the system's energy in the range E to $E + dE$ when the temperature is prescribed is $Q(E)\,dE$, this means that the distribution function $Q(E)$ must be very strongly peaked about $E = \bar{E}$, as in Fig. 1.4. The figure shows the distribution to have some width, δE. The energy $\bar{E}\,(=U)$, being an extensive thermodynamic function, is proportional to the size of the system. We may conveniently take the number of its molecules, or total number of its mechanical degrees of freedom, to be a dimensionless measure of the system's size. Call this number N. It will typically be of the order of magnitude of Avogadro's number, say 10^{22} to 10^{24}. Then the mean energy $\bar{E}$ will be of that order of magnitude; i.e., it will be of order $\bar{\varepsilon}N$, where the intensive $\bar{\varepsilon}$ is the energy per molecule or per degree of freedom. With such a measure N of the system size, the typical energy fluctuations δE (Fig. 1.4) are only of order $\sqrt{N}$; i.e., they

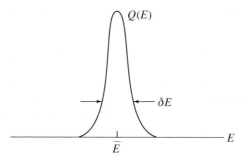

Fig. 1.4

1.3 The partition function and statistical thermodynamics

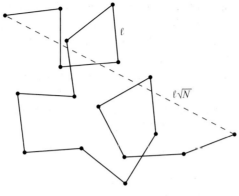

Fig. 1.5

are typically smaller than $\bar{E}$ itself by a factor of 10^{-11} or 10^{-12}. We can thus see why such fluctuations are thermodynamically inconsequential, and, indeed, why they are not even observable by ordinary macroscopic means.

The $\sqrt{N}$ law for the fluctuations is the same as that in a random walk. Suppose a walker takes a large number N of steps, each step of length ℓ but in a random direction, uncorrelated with that of the preceding step. How far, on average, has the walker gotten from the starting point after those N steps? (See Fig. 1.5.) Answer: $\ell\sqrt{N}$. In our system of N molecules or N degrees of freedom, each has, at any instant, some energy that differs more or less from the average (that is the analog of the step-length ℓ), but the difference for each molecule or each degree of freedom is randomly positive or negative (the analog of the random step directions), with the result that the net departure of the total energy of the whole system from its average is of order $\sqrt{N}$.

Except for the normalization constant Z^{-1}, the energy distribution function $Q(E)$ given by (1.22) is just the integrand in the formula (1.21) for the partition function. (Indeed, substituting (1.22) into (1.21) confirms that the distribution $Q(E)$ in (1.22) is properly normalized: its integral over the full energy range is 1.) We know this integrand, then, to be of reasonable magnitude only within a relatively narrow range of energies – perhaps a few multiples of δE (Fig. 1.4) – about $\bar{E}$, and to fall rapidly to negligibly small values as E departs from $\bar{E}$. The integral (1.21), then, will be essentially the integrand evaluated at its maximum, $Q(\bar{E}) = \exp(-\bar{E}/kT)W(\bar{E})$, multiplied by some energy interval, say ΔE, that is of the order of the peak-width δE:

$$Z = e^{-\bar{E}/kT} W(\bar{E}) \Delta E. \qquad (1.23)$$

For example, $Q(E)$ might be approximately the normalized Gaussian distribution

$$Q(E) = Z^{-1} e^{-E/kT} W(E) = \frac{1}{\sqrt{\pi} \delta E} e^{-(E-\bar{E})^2/(\delta E)^2}, \quad (1.24)$$

which has a peak at $E = \bar{E}$ of width of order δE. [The $1/e$-width, i.e., the width of the peak where it has fallen to the fraction $1/e$ of the value at its maximum, is $2\,\delta E$. Note that $\int_{-\infty}^{\infty} e^{-x^2}\,dx = \sqrt{\pi}$.] This $Q(E)$ is normalized over the interval $-\infty$ to ∞ instead of from some finite E_0 to ∞, but it falls off so rapidly as soon as E departs from $\bar{E}$ by more than a few multiples of δE that the integral (1.21) for Z might just as well be extended to $-\infty$.

We see from (1.24) that $Z^{-1} \exp(-\bar{E}/kT) W(\bar{E}) = (\sqrt{\pi}\delta E)^{-1}$, or

$$Z = e^{-\bar{E}/kT} W(\bar{E}) \sqrt{\pi} \delta E, \quad (1.25)$$

essentially as asserted in (1.23).

Exercise (1.2).
(a) Show that the energy probability distribution function $Q(E)$ given by Eq. (1.24) is normalized over the interval $E = -\infty$ to ∞; i.e., that

$$\int_{-\infty}^{\infty} Q(E)\,dE = 1.$$

(b) For the $Q(E)$ given by Eq. (1.24), plot $\bar{E} Q(E)$ as a function of $E/\bar{E}$ for $\delta E/\bar{E} = 1/5, 1/10$, and $1/20$; and then try to imagine (!) what the corresponding plot would look like if $\delta E/\bar{E}$ had the more realistic value 10^{-11}.

Solution.

(a)
$$\int_{-\infty}^{\infty} Q(E)\,dE = \frac{1}{\sqrt{\pi}\delta E} \int_{-\infty}^{\infty} e^{-(E-\bar{E})^2/(\delta E)^2}\,dE$$
$$= \frac{1}{\sqrt{\pi}} \int_{-\infty}^{\infty} e^{-x^2}\,dx \quad [(E-\bar{E})/\delta E = x]$$
$$= 1.$$

[The value of the integral over x is as quoted beneath Eq. (1.24).]

(b) Rewrite Eq. (1.24) as

$$\bar{E} Q(E) = \frac{1}{\sqrt{\pi}\delta E/\bar{E}} e^{-\left(\frac{E/\bar{E}-1}{\delta E/\bar{E}}\right)^2}$$

1.3 The partition function and statistical thermodynamics

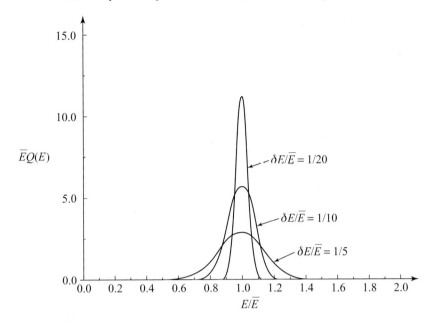

and plot this as a function of $E/\bar{E}$, with $\delta E/\bar{E}$ fixed, in turn, at $1/5$, $1/10$, and $1/20$. The plots are in the accompanying figure. The peak value of $\bar{E}Q(E)$ for $\delta E/\bar{E} = 10^{-11}$ would be greater by a factor of 10^{10} than the peak value $10/\sqrt{\pi} = 5.64$ seen in the figure for $\delta E/\bar{E} = 1/10$, while the width of the peak would be less by the same factor; so that on the scale of this plot the peak for $\delta E/\bar{E} = 10^{-11}$ would appear to be infinitely sharp, having a practically infinite height and no perceptible width.

With $\bar{E}$ identified with the thermodynamic energy U at the given T, with the relation $A = -kT \ln Z$ from (1.16), and with $A = U - TS$, Eq. (1.23) gives for the entropy,

$$S = (U - A)/T = (\bar{E} + kT \ln Z)/T$$
$$= \{\bar{E} - \bar{E} + kT \ln[W(\bar{E})\Delta E]\}/T$$
$$= k \ln[W(U)\Delta E]. \qquad (1.26)$$

The argument of the logarithm, $W(U)\Delta E$, is the density of states $W(U)$ at the energy U multiplied by what is essentially the range of energies over which the energy of the system might with reasonable probability fluctuate when the temperature is prescribed. Thus, $W(U)\Delta E$ may be interpreted as the number of states to which the system has access when its temperature is given (see

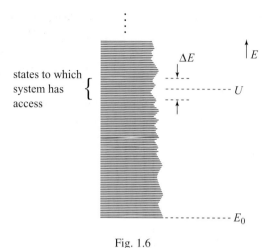

Fig. 1.6

Fig. 1.6); and the entropy may now be understood as a logarithmic measure of that number of accessible states. That is the microscopic interpretation of entropy.

Since the entropy is extensive, the right-hand side of (1.26) must be of order N, the number of molecules or number of degrees of freedom, in the language we used earlier. But by that measure ΔE is only of order $\sqrt{N}$, we know, and so it only contributes to S a piece of order $\ln N$, which is negligibly small compared with N when N is large. It must then be that $W(U)\Delta E$ is of the order of the *exponential* of a quantity of order N and that almost all of that comes from the density of states W. The only reason for retaining the factor ΔE in the argument of the logarithm is to keep that argument dimensionless – $W\Delta E$ is a number of states – but that factor is otherwise of no significance in the formula (1.26). The number of accessible states $W\Delta E$ being of the order of e *raised to a power* of about 10^{23} means that it is almost unimaginably large – probably the largest number we shall ever in our lives have reason to contemplate.

The formula $S = k\ln(W\Delta E)$ (or the dimensionally improper $S = k\ln W$, as it is more often written) is probably the most famous and most quoted formula of statistical mechanics, although not the most frequently used in practice. It is due to Boltzmann, whose name has been deservedly given to the proportionality constant k. Since this formula preceded quantum mechanics, W could not originally have been the density of quantum states, as it is here, but was rather its analog in classical mechanics.

1.3 The partition function and statistical thermodynamics

We noted earlier that the density of states $W(E)$ depends also on the volume and composition of the system, so that $W(U)$ in Eq. (1.26) for the entropy, expressed more fully, is $W(U, V, N_1, N_2, \ldots)$. Then (1.26) (or more simply the dimensionally improper $S = k \ln W$) directly yields the entropy as a function of those variables: $S(U, V, N_1, N_2, \ldots)$. This, like $A(T, V, N_1, N_2, \ldots)$, is a thermodynamic potential; all other thermodynamic functions are derivable from $S(U, V, N_1, N_2, \ldots)$ by differentiations alone, no integrations being necessary. For example, the absolute temperature T, pressure p, and chemical potentials μ_i are obtained from the thermodynamic identities

$$T = \left(\frac{\partial S}{\partial U}\right)^{-1}_{V, N_1, N_2, \ldots} \tag{1.27}$$

$$p = T\left(\frac{\partial S}{\partial V}\right)_{U, N_1, N_2, \ldots} \tag{1.28}$$

$$\mu_1 = -T\left(\frac{\partial S}{\partial N_1}\right)_{U, V, N_2, N_3, \ldots}, \quad \text{etc.} \tag{1.29}$$

Thus, $S = k \ln(W \Delta E)$ (or $S = k \ln W$) is an alternative to $A = -kT \ln Z$ as a route to the thermodynamic functions. The density of states W (or more properly $W \Delta E$) is also called a partition function – the *microcanonical* partition function, to distinguish it from Z, which is called the *canonical* partition function. Equation (1.21) showed the latter to be the Laplace transform of the former. Much later, in Chapter 8, we shall encounter still another partition function.

In the next chapter we shall evaluate the partition function Z, and from it the thermodynamic functions, for the simplest of all systems, the ideal gas.

2
The ideal gas

2.1 Boltzmann statistics

As a first step in formulating and evaluating the partition function Z for an ideal gas (a gas of non-interacting molecules) by the general formula (1.12), we must specify the states i of the whole system and their energies E_i. Suppose the gas comprises N molecules. Let n_1 be symbolic for all the quantum numbers (translational, vibrational, rotational, ...) of molecule number 1, similarly n_2 for those of molecule number 2, etc., through n_N for the Nth molecule. Each of these n's is just shorthand for a collection of many quantum numbers when the molecule to which it refers is of any complexity. At the very least each such n is a composite of three translational quantum numbers; that simplest case occurs when the molecule is monatomic (a rare gas atom, for example) and the temperature is low enough so that we may ignore all but the ground electronic state. In such a case we may model the atom as a structureless point particle with only its three translational degrees of freedom.

Tentatively, we might suppose that to specify a state of the whole system we have merely to specify the state of each molecule; i.e., the sets of quantum numbers $n_1, \ldots, n_N$; and the energy E in that state would then just be the sum $\varepsilon_{n_1} + \cdots + \varepsilon_{n_N}$ of the energies $\varepsilon_{n_1}, \varepsilon_{n_2}, \ldots$ of the individual molecules when they are in those specified states. The sum (1.12) over the states i of the whole system would then be a multiple sum over all the possible assignments of all the quantum numbers in the quantum-number set n_1 for molecule 1, over all those in the set n_2, etc.; so, symbolically, that sum would now be

$$\sum_{n_1} \sum_{n_2} \cdots \sum_{n_N} e^{-(\varepsilon_{n_1} + \varepsilon_{n_2} + \cdots + \varepsilon_{n_N})/kT}. \tag{2.1}$$

Since the exponential of a sum is the product of exponentials, the summand in (2.1) is the product of factors one of which depends only on the summation variables n_1, another only on the summation variables n_2, etc. The multiple sum

2.1 Boltzmann statistics

is then just the product of the separate, independent sums:

$$\zeta_1 \zeta_2 \ldots \zeta_N \tag{2.2}$$

where

$$\zeta_1 = \sum_{n_1} e^{-\varepsilon_{n_1}/kT}, \quad \zeta_2 = \sum_{n_2} e^{-\varepsilon_{n_2}/kT}, \text{ etc.} \tag{2.3}$$

Exercise (2.1). Illustrate this decomposition of a multiple sum with the simple example of a double sum, in which the summand is a product of two factors, one of which depends on only one of the two summation variables and the other of which depends only on the other variable.

Solution. The double sum is of the form

$$\sum_k \sum_\ell f_k g_\ell = \sum_k \left(\sum_\ell f_k g_\ell \right) \quad \text{(summing first over } \ell, \text{ then over } k\text{)}$$

$$= \sum_k \left(f_k \sum_\ell g_\ell \right) \quad \text{(because } f_k \text{ is independent of } \ell\text{)}$$

$$= \left(\sum_\ell g_\ell \right) \sum_k f_k \quad \text{(because } \sum_\ell g_\ell \text{ is independent of } k\text{)},$$

which is the decomposition into a product of individual sums that we wished to illustrate. As a simple special case of this example, assume k and ℓ each to take on the three values 1, 2, 3. Then

$$\sum_{k=1}^{3} \sum_{\ell=1}^{3} f_k g_\ell = f_1 g_1 + f_1 g_2 + f_1 g_3 + f_2 g_1 + f_2 g_2$$

$$+ f_2 g_3 + f_3 g_1 + f_3 g_2 + f_3 g_3$$

while

$$\left(\sum_{k=1}^{3} f_k \right) \left(\sum_{\ell=1}^{3} g_\ell \right) = (f_1 + f_2 + f_3)(g_1 + g_2 + g_3),$$

which, when multiplied out, yields the same nine terms as in the preceding expression.

According to (2.3), each factor ζ in the product (2.2) is a single-molecule partition function: the sum, over all the states of a single molecule, of the

Boltzmann factor associated with that state of that molecule. But many, maybe all, of the N molecules in the system are of the same kind. Suppose the first N_1 are of one kind, the next N_2 are of another kind, etc., with

$$N_1 + N_2 + N_3 + \cdots = N. \tag{2.4}$$

Then the first N_1 factors ζ in (2.2) have the common value z_1, say, the next N_2 factors have the common value z_2, etc. The product (2.2) is then

$$z_1{}^{N_1} z_2{}^{N_2} z_3{}^{N_3} \cdots, \tag{2.5}$$

where now z_1 is the single-molecule partition function of a molecule of species 1, z_2 that of a molecule of species 2, etc. The only difference between a ζ and a z is that the subscript on a ζ identifies the molecule while that on a z identifies the species.

We have been tentatively supposing that for this gas of non-interacting molecules we specify a single state i of the whole system by specifying the values of the quantum numbers $n_1, \ldots, n_N$ of the N molecules that compose the system, and that has led to the reduction of the multiple sum in the partition function (1.12) to the product, (2.5), of single-molecule partition functions. But now there is a difficulty. Suppose the molecules numbered 1 and 2 are identical; i.e., that they are two molecules of the same species. The multiple sum (2.1) includes terms in which the quantum numbers n_1 of molecule 1 take on one set of values and the quantum numbers n_2 of molecule 2 take on another, but also includes terms in which molecule 2 is assigned the set previously assigned to molecule 1 and vice versa. But since molecules 1 and 2 are identical, these are not truly distinct quantum states of the pair; they are the same state, merely with two different designations. A state of the *pair* is obtained by specifying that *one* molecule of the pair – either one – is in state n_1 and that the *other* – whichever it is – is in state n_2. Thus, the sum (2.1) overcounts the distinct states i that we were supposed to be summing over in (1.12): there are identical terms in (2.1) that occur many times over, merely under different names, when they should occur only once. The total redundancy, taking account of there being N_1 identical molecules of one kind, N_2 of another, etc., might seem at first sight to be just the numbers of permutations of identical particles; i.e., the product of factorials $N_1! \, N_2! \cdots$; because that, we might at first have supposed, is the number of times the term corresponding to state i in (1.12) is repeated, merely under different names (that is, by permutation of identical particles), in the sum (2.1). Thus, we might have supposed that the partition function Z of our ideal gas was really

$$Z = \frac{1}{N_1! N_2! \cdots} z_1{}^{N_1} z_2{}^{N_2} \cdots. \tag{2.6}$$

2.1 Boltzmann statistics

But now return to the original uncorrected multiple sum (2.1) and suppose again that molecules 1 and 2 are identical. There is a term in (2.1) in which the quantum numbers n_1 of molecule 1 take on one set of values and the quantum numbers n_2 of molecule 2 take on the *same* set of values. That term occurs only once in (2.1), not twice. Such states are therefore not overcounted in (2.1). Thus, (2.1) does not overcount the states by *all* permutations of identical particles: only by those in which identical particles are in different states, not by those in which the identical particles are in the same state. Therefore, dividing (2.1) or the equivalent (2.5) by $N_1!N_2!\cdots$ to correct for the overcounting is an *over*correction. The result is that (2.6) is an *approximation* for the ideal gas; it is not exact. The approximation (2.6) is called the *Boltzmann statistics* of the ideal gas.

When is the overcorrection serious and when is it negligible? If in the multiple sum (2.1) there are relatively few terms in which identical particles are in the same state, then the overcorrection, which treats only those terms incorrectly, will be of little consequence and the partition function Z will be given nearly exactly by (2.6). But if terms in which identical particles are in the same state occur prominently in (2.1), then (2.6) may be seriously in error. In Fig. 2.1 is shown a schematic energy-level diagram of the single-molecule energies ε, indicating a range of energy, equal to the thermal energy kT, above the ground-state energy ε_0. The levels in Fig. 2.1 are not nearly as dense as those in Figs. 1.3 and 1.6 of Chapter 1, which are the energy levels of a whole macroscopic system rather than of a single molecule, as here; but they are nevertheless very dense here, too, because the states include the translational states of a particle in a box of macroscopic size, and translational energy-level spacings decrease with increasing box size, as we remarked earlier (§1.3). According to the Boltzmann distribution law (1.1), only those levels within kT, or within a few kT, of the

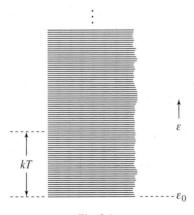

Fig. 2.1

ground state will be significantly populated. Thus, whether identical molecules' being in the same state is or is not important depends entirely on whether the number of states in the range kT above ε_0 shown in Fig. 2.1 is or is not much larger than the number of molecules. If there are many more states in that range than there are molecules, then multiple occupancy would in any case be so rare an event that treating it incorrectly would lead to negligible error, and the Boltzmann statistics, i.e., the approximation (2.6), would be essentially exact.

If, on the other hand, the number of such levels is not much greater than the number of molecules, multiple occupancy becomes a significant possibility and (2.6) could be seriously in error. In that event one must distinguish the particles as *bosons* or *fermions*. The former, being spinless or consisting of an even number of spin-$\frac{1}{2}$ particles, are not restricted in their occupancy of states: any number can be in the same state. The latter, being of spin $\frac{1}{2}$ or consisting of an odd number of spin-$\frac{1}{2}$ particles, are subject to the Pauli exclusion principle. As one knows from quantum mechanics, no two of such identical particles can be in the same state. (We include the spin quantum number among the quantum numbers that define the state.) Thus, when the possibility of multiple occupancy of states has to be considered, so that the Boltzmann statistics is no longer an adequate approximation, the nature of the required corrections to it depends on the nature of the particles that compose the ideal gas in question: for bosons the Boltzmann statistics must be replaced by the correct *Bose–Einstein* statistics while for fermions it is the *Fermi–Dirac* statistics that must be used.

Low density (i.e., for a given number of particles, a large volume), large mass of the molecules, and high temperature, all favor the Boltzmann statistics: the former two because with large enough volume or high enough mass the energy-level spacing can be made so small that there are many more states within kT of ε_0 than there are particles, so that multiple occupancy of states ceases to be a consideration, and the latter then for the same reason. In practically every case of interest in physical chemistry that condition is satisfied and the Boltzmann statistics is a good – indeed, an outstandingly good – approximation. The approximation fails, however, for an electron gas at the typical density of the valence electrons in a metal, because of the exceptionally small mass of the electron. The Fermi–Dirac statistics is then essential. The condition fails also for helium at densities typical of that of the liquid because of the exceptionally low temperatures, only a few kelvin, at which one would be interested in it. There are hardly any other exceptions to the rule that (2.6) is a very good approximation to the partition function of a gas of non-interacting molecules.

Sometimes the Boltzmann statistics is referred to as "classical," as distinct from the Bose–Einstein and Fermi–Dirac statistics, which are called "quantum" statistics. The latter are properly called quantum statistics, but to refer to the former as classical is very misleading. The single-molecule partition functions z (or ζ), calculated by (2.3), may be as quantum mechanical as we wish; they may be calculated as sums over distinct quantum states. To be sure, we shall later take the classical limits of the translational parts of the partition functions, and often also of the rotational parts, but only rarely shall we treat the molecular vibrations classically and never the electronic degrees of freedom. Even for the rotations the quantum mechanical aspects will sometimes be of interest.

Those cases, such as the electron gas or dense helium at low temperatures, where the Fermi–Dirac or Bose–Einstein statistics are essential, although rarely arising in physical chemistry, are nevertheless interesting and important. We shall treat them in Chapter 8.

2.2 Translational partition function

The translational motion of a molecule, i.e., the motion of the molecule's center-of-mass, is rigorously separable from its other degrees of freedom, in both classical and quantum mechanics. Among the quantum numbers that we have before symbolized collectively by the single symbol n in the formula

$$z = \sum_n e^{-\varepsilon_n/kT} \tag{2.7}$$

for the single-molecule partition function z, there are three that define the translational state. If we symbolize these separately by n_{trans}, and the remaining ones, which refer to the internal degrees of freedom (rotational, vibrational, and electronic), as n_{int}, we then have, more explicitly,

$$z = \sum_{n_{\text{trans}}} \sum_{n_{\text{int}}} e^{-(\varepsilon_{n_{\text{trans}}} + \varepsilon_{n_{\text{int}}})/kT}, \tag{2.8}$$

where account has now been taken of the separability: $\varepsilon_n = \varepsilon_{n_{\text{trans}}} + \varepsilon_{n_{\text{int}}}$. The summand in (2.8) is again a product of factors that depend on different ones of the summation indices, so the multiple sum is once more a product of separate sums:

$$z = z_{\text{trans}} z_{\text{int}}, \tag{2.9}$$

where z_{trans} and z_{int} are the partition functions of the translational and internal degrees of freedom of a single molecule,

$$z_{\text{trans}} = \sum_{n_{\text{trans}}} e^{-\varepsilon_{n_{\text{trans}}}/kT}, \quad z_{\text{int}} = \sum_{n_{\text{int}}} e^{-\varepsilon_{n_{\text{int}}}/kT}. \tag{2.10}$$

In the present section of this chapter we evaluate z_{trans}; subsequent sections deal with z_{int}.

The three translational degrees of freedom of the molecule are independent of each other, so by the same process as that above we may see that z_{trans} factorizes further into the product of three factors, one for each translational degree of freedom. Imagine the volume to which the gas is confined to be a rectangular box with sides of lengths L_1, L_2, and L_3 (so that the volume $V = L_1 L_2 L_3$), and let z_{trans_1}, for example, be the factor in z_{trans} that refers to the component of the particle's motion in the direction parallel to the sides of length L_1. Let the molecule's mass be m. The energy levels of a particle of mass m confined to a one-dimensional box of length L_1 are

$$\varepsilon_n = \frac{h^2 n^2}{8m L_1^2} \quad (n = 1, 2, 3, \ldots), \quad (2.11)$$

where now the symbol n is being used for this single quantum number, and where h, as usual, is Planck's constant.

We then have

$$z_{\text{trans}_1} = \sum_{n=1}^{\infty} e^{-h^2 n^2 / 8m L_1^2 kT}. \quad (2.12)$$

The significant contributions to this sum come from values of n up to about $n_{\max} = \sqrt{8mkT}\, L_1/h$; for $n \gg n_{\max}$ the exponential falls off rapidly with increasing n and those further terms are then negligible. With the realistic magnitudes $m = 10^{-22}$ g, $T = 300$ K, and $L_1 = 1$ cm, this $n_{\max}$ is about 10^9, which is enormous. Now, from one term in the sum to the next the fractional decrease in the summand is about

$$-\frac{1}{e^{-h^2 n^2 / 8m L_1^2 kT}} \frac{d}{dn} e^{-h^2 n^2 / 8m L_1^2 kT} \quad (2.13)$$

(the derivative being an estimate of the change in the summand that occurs when the summation index changes by 1), which is

$$-\frac{d}{dn} \ln e^{-h^2 n^2 / 8m L_1^2 kT} = 2h^2 n / 8m L_1^2 kT = 2n/n_{\max}^2. \quad (2.14)$$

Thus, for small n ($n = 1, 2, 3, \ldots$) the fractional decrease from one term to the next is negligibly small; indeed, for such small n the summand, which is $\exp(-n/n_{\max})^2$, is practically just the constant 1; while even for n of the order of $n_{\max}$ this fractional decrease is of order $1/n_{\max}$, which is still minute. The fractional decrease does not become significant until n becomes of order $n_{\max}^2 \gg n_{\max}$, but by then the terms have long since ceased to contribute significantly to the sum. In short, over the whole significant range of n the terms of the sum change only slightly from one to the next. That is exactly the condition under

2.2 Translational partition function

which the sum may be very closely approximated by an integral, so for all practical purposes

$$z_{\text{trans}_1} = \int_1^\infty e^{-(n/n_{\max})^2} dn$$

$$= n_{\max} \int_{1/n_{\max}}^\infty e^{-x^2} dx$$

$$\approx n_{\max} \int_0^\infty e^{-x^2} dx \qquad (2.15)$$

(because, as we saw, $n_{\max}$ is huge). The integral is $\frac{1}{2}\sqrt{\pi}$ and $n_{\max}$ was $\sqrt{8mkT}\, L_1/h$, so we now have, to an approximation that is so good as to be virtually exact,

$$z_{\text{trans}_1} = \sqrt{2\pi mkT}\, L_1/h. \qquad (2.16)$$

The whole translational partition function is the product of three such factors, one with L_1, one with L_2, and one with L_3. Since $L_1 L_2 L_3$ is the volume V,

$$z_{\text{trans}} = \left(\frac{2\pi mkT}{h^2}\right)^{3/2} V. \qquad (2.17)$$

Replacing the sum (2.12) by an integral, which we saw was well justified, was equivalent to ignoring the quantization of the translational energy levels; i.e., to taking the classical limit of the quantum mechanical z_{trans}. What justified it was the enormous size of $n_{\max}$; i.e., of $\sqrt{8mkT}\, L_1/h$. It is generally true that the classical limit of a quantum mechanical formula is obtained in the limit of a large mass m, or of a high temperature T, or of a large spatial extent L, or even, formally, as the limit "$h \to 0$." We shall see more examples of this later.

The factor z_{int} in the single-molecule partition function z depends only on the internal degrees of freedom of the molecule and so is independent of the volume V. The entire volume dependence in the partition function Z of the whole system, as given by (2.6), comes only from the factors V in the translational parts z_{trans} of the individual $z_1, z_2, \ldots$ in (2.6). The whole V dependence of Z is then in a factor $V^{N_1+N_2+\cdots} = V^N$, with $N = N_1 + N_2 + \cdots$ again, as before [Eq. (2.4)], the total number of molecules in the system irrespective of their kind. Once we take the logarithm of Z to obtain the Helmholtz free energy A [Eq. (1.16)], and then differentiate with respect to V at fixed $T, N_1, N_2, \ldots$ to obtain the pressure p [Eq. (1.19)], only the V-dependent factor V^N in Z survives, and yields

$$p = [\partial(kT \ln V^N)/\partial V]_T = NkT/V. \qquad (2.18)$$

Recall that $Nk = nR$, with n the number of moles and R the gas constant. Thus, (2.18) is the familiar ideal-gas law, which we might have expected for this model of non-interacting molecules. We see that the pressure does not depend on the internal degrees of freedom of the molecules, on their masses, or on anything else about them except their total number.

Note at this point that if the additive $T\phi(V, N_1, N_2, \ldots)$ in (1.15) of the preceding chapter really did depend on V there would have been an additional term in p in (2.18). It is our prior knowledge of the absence of any such term for a dilute gas that tells us that the ϕ in (1.15) in fact does not depend on V; for any given system (given $N_1, N_2, \ldots$) it is some constant, which, we saw there, is connected with the arbitrary zero of entropy. We follow the usual convention of setting $\phi = 0$.

Because the thermodynamic properties of a system, from (1.16)–(1.20), come from the logarithm of the partition function Z, and because the logarithm of a product is a sum of logarithms, each factor in Z makes its own additive contribution to A, S, U, p, μ_i, etc. Because of the factorizations (2.6) and (2.9), then, the translational degrees of freedom contribute additively to the thermodynamic functions and those contributions may be calculated from the z_{trans} of (2.17). For example, from the factorizations (2.6) and (2.9) and from z_{trans} in (2.17), the translational degrees of freedom contribute to Z, multiplicatively, the temperature dependence $T^{3N/2}$ with $N = N_1 + N_2 + \cdots$ again. Therefore, from (1.16), they contribute $(-3NkT/2) \ln T$ to A, and so from the Gibbs–Helmholtz equation (1.14) they contribute

$$\begin{aligned} E_{\text{trans}} &= -(3Nk/2)\,\mathrm{d}\ln T/\mathrm{d}(1/T) \\ &= (3Nk/2)\,\mathrm{d}\ln(1/T)/\mathrm{d}(1/T) \\ &= (3Nk/2)/(1/T) \\ &= \frac{3}{2}NkT \end{aligned} \qquad (2.19)$$

to the total thermodynamic energy U. This is in accord with what one learns from the kinetic theory of gases: the average translational energy per molecule is $\frac{3}{2}kT$ (that is, $\frac{1}{2}kT$ per degree of freedom) whatever the mass or internal structure of the molecule.

One learns in thermodynamics that the constant-volume heat capacity $C_V = (\partial U/\partial T)_V$. Here, then, the temperature derivative of E_{trans} gives the corresponding contribution, C_{trans}, to the heat capacity:

$$C_{\text{trans}} = \frac{3}{2}Nk, \qquad (2.20)$$

or $\frac{3}{2}R$ per mole, or $\frac{1}{2}R$ per mole per degree of freedom.

If the gas consists of a single component, so that all N molecules are of one kind, then the full translational contribution to A (with the $1/N!$ in (2.6)

2.2 Translational partition function

incorporated) is now

$$A_{\text{trans}} = -kT \left(\ln z_{\text{trans}}^N - \ln N! \right)$$
$$= -kT \left\{ N \ln \left[\left(\frac{2\pi mkT}{h^2} \right)^{3/2} V \right] - \ln N! \right\}. \quad (2.21)$$

An important approximation – the *Stirling formula* – allows us to express the term $\ln N!$ in a more useful way.

The number N is very large, being typically of the order of 10^{23}. The factorial of any large number N is given accurately by

$$N! \sim N^N e^{-N} \sqrt{2\pi N}, \quad (2.22)$$

which is the Stirling formula. [The symbol $\sim$ means "asymptotically equal to." $a \sim b$ in some limit means that the limiting value of a/b is 1. In (2.22) the contemplated limit is $N \to \infty$.] At large N the fractional error one incurs in approximating $N!$ by the right-hand side of (2.22) is $1/(12N)$. Even for N as small as 8, then, the approximation is already good to about 1%; indeed, $8! = 40{,}320$ while $8^8 e^{-8} \sqrt{16\pi} \simeq 39{,}902$, with an error of about 400 out of 40,000, or 1%. With N as large as 10^{23} the Stirling approximation is virtually exact.

With $N!$ approximated by the right-hand side of (2.22), the formula (2.21) for A_{trans} becomes

$$A_{\text{trans}} = -kT \left\{ N \ln \left[\left(\frac{2\pi mkT}{h^2} \right)^{3/2} \frac{V}{N} e \right] - \ln \sqrt{2\pi N} \right\}. \quad (2.23)$$

But $\ln \sqrt{2\pi N}$ is of the order of magnitude of $\ln N$, which is sub-extensive; i.e., smaller in order of magnitude than the leading term, which is of order N. (Note that V/N, hence the entire argument of the first logarithm, is intensive.) We may therefore neglect the $\ln \sqrt{2\pi N}$ term, with the result

$$A_{\text{trans}} = -NkT \ln \left[\left(\frac{2\pi mkT}{h^2} \right)^{3/2} \frac{V}{N} e \right], \quad (2.24)$$

which is the final formula for A_{trans}.

The corresponding contribution to the entropy, S_{trans}, now follows from (1.17), or, equivalently, from (1.18), since E_{trans} is given in (2.19). The second route is here the simpler:

$$S_{\text{trans}} = (E_{\text{trans}} - A_{\text{trans}})/T$$
$$= Nk \ln \left[\left(\frac{2\pi mkT}{h^2} \right)^{3/2} \frac{V}{N} e^{5/2} \right]. \quad (2.25)$$

This is S_{trans} for given N, V, T. An equivalent formula for use when N, p, T are given is obtained by replacing V/N by kT/p, from the ideal-gas law (2.18); thus,

$$S_{\text{trans}} = Nk \ln\left[\left(\frac{2\pi mkT}{h^2}\right)^{3/2} \frac{kT}{p} e^{5/2}\right]. \tag{2.26}$$

This is the *Sackur–Tetrode* equation.

Exercise (2.2). Calculate the entropy change ΔS_{mix} (the "entropy of mixing") when two different ideal gases, one consisting of N_1 molecules of one species, at temperature T and pressure p, and the other consisting of N_2 molecules of a second species, at the same T and p, are mixed, and the mixture is again at that same T and p.

Solution. The contributions of the internal degrees of freedom of the molecules to A and then to S are the same before and after mixing and may thus be ignored; the whole of ΔS_{mix} comes from S_{trans} alone. For the mixture, if the masses of the molecules of the two species are m_1 and m_2,

$$A_{\text{trans,mixture}} = -kT \ln\left[\frac{1}{N_1! N_2!}\left(\frac{2\pi m_1 kT}{h^2}\right)^{3N_1/2}\left(\frac{2\pi m_2 kT}{h^2}\right)^{3N_2/2} V^{N_1+N_2}\right].$$

Replace each factorial by (2.22) (ignoring the factors $\sqrt{2\pi N_i}$, the logarithms of which are negligible):

$$A_{\text{trans,mixture}} = -kT \ln\left[\left(\frac{2\pi m_1 kT}{h^2}\right)^{3N_1/2}\left(\frac{2\pi m_2 kT}{h^2}\right)^{3N_2/2}\right.$$
$$\left.\times \left(\frac{V}{N_1}\right)^{N_1}\left(\frac{V}{N_2}\right)^{N_2} e^{N_1+N_2}\right].$$

From (2.19) the total translational energy of the molecules in the mixture is $E_{\text{trans,mixture}} = \frac{3}{2}(N_1 + N_2)kT$, so the translational entropy of the mixture is

$$S_{\text{trans,mixture}} = (E_{\text{trans,mixture}} - A_{\text{trans,mixture}})/T$$
$$= k \ln\left[\left(\frac{2\pi m_1 kT}{h^2}\right)^{3N_1/2}\left(\frac{2\pi m_2 kT}{h^2}\right)^{3N_2/2}\right.$$
$$\left.\times \left(\frac{V}{N_1}\right)^{N_1}\left(\frac{V}{N_2}\right)^{N_2} e^{5(N_1+N_2)/2}\right].$$

Replace V by $(N_1 + N_2)kT/p$ and introduce the mole fractions $x_1 = N_1/(N_1 + N_2)$ and $x_2 = N_2/(N_1 + N_2)$:

$$S_{\text{trans,mixture}} = N_1 k \ln\left[\left(\frac{2\pi m_1 kT}{h^2}\right)^{3/2} \frac{kT}{p} e^{5/2}\right]$$
$$+ N_2 k \ln\left[\left(\frac{2\pi m_2 kT}{h^2}\right)^{3/2} \frac{kT}{p} e^{5/2}\right]$$
$$+ N_1 k \ln(1/x_1) + N_2 k \ln(1/x_2).$$

But the first two of these logarithmic terms, by (2.26), are the translational entropies S_{trans_1} and S_{trans_2} of the separate gases before mixing. Then

$$\Delta S_{\text{mix}} = S_{\text{trans,mixture}} - \left(S_{\text{trans}_1} + S_{\text{trans}_2}\right)$$
$$= k[N_1 \ln(1/x_1) + N_2 \ln(1/x_2)]$$
$$= (N_1 + N_2)k[x_1 \ln(1/x_1) + x_2 \ln(1/x_2)].$$

Since $(N_1 + N_2)k$ is the same as $(n_1 + n_2)R$, this is the result familiar in thermodynamics.

2.3 Vibrational partition function

The formal single-molecule partition function z, according to (2.9), has as separate factors the partition functions z_{trans} and z_{int} associated with the molecule's translational and internal degrees of freedom. To the extent that the vibrational and rotational degrees of freedom may be separable from the electronic (the Born–Oppenheimer approximation of quantum mechanics), and the vibrational from the rotational (a further approximation), and that the temperature is low enough so that we need consider only a single electronic state – the ground state – of the molecule, z_{int} will factorize further into the product of vibrational and rotational partition functions:

$$z_{\text{int}} \simeq z_{\text{vib}} z_{\text{rot}}. \tag{2.27}$$

Finally, if the temperature is low enough so that the vibrations are not highly excited, they will be at least approximately harmonic and separable into independent normal modes. Then z_{vib} will factorize further into the product of independent vibrational partition functions z_{vib_1}, z_{vib_2}, etc., each associated with a single normal mode:

$$z_{\text{vib}} \simeq z_{\text{vib}_1} z_{\text{vib}_2} \cdots. \tag{2.28}$$

Now, for simplicity, let us change the notation slightly and let z_{vib} stand for any one of the factors z_{vib_i} on the right-hand side of (2.28), instead of for the product of all of them. If this vibration is of frequency ν and if, as is conventional, we choose the zero of energy to be at the bottom of the parabolic potential, then the energy levels are $\varepsilon_v = (v + \frac{1}{2})h\nu$ with $v = 0, 1, 2, \ldots$. Therefore we have

$$z_{vib} = \sum_{v=0}^{\infty} e^{-(v+\frac{1}{2})h\nu/kT}$$

$$= e^{-\frac{1}{2}h\nu/kT} \sum_{v=0}^{\infty} \left(e^{-h\nu/kT}\right)^v$$

$$= \frac{e^{-\frac{1}{2}h\nu/kT}}{1 - e^{-h\nu/kT}}$$

$$= \frac{1}{e^{\frac{1}{2}h\nu/kT} - e^{-\frac{1}{2}h\nu/kT}}. \tag{2.29}$$

In going from the second line to the third we made use of the formula for the sum of an infinite geometric series, $1 + x + x^2 + x^3 + \cdots = 1/(1-x)$.

There is a temperature θ_{vib} that characterizes this vibrational mode,

$$\theta_{vib} = h\nu/k, \tag{2.30}$$

such that, with respect to this vibration, a temperature T would be very low or very high according to whether it is much below or much above θ_{vib}. For low temperatures, $T \ll \theta_{vib}$, the second exponential in the denominator of (2.29) is negligible compared with the first, and

$$z_{vib} \sim e^{-\theta/2T} \quad (T \ll \theta), \tag{2.31}$$

where θ is now just shorthand for the θ_{vib} given by (2.30). For high temperatures, $T \gg \theta_{vib}$, we may expand the exponentials as $\exp x = 1 + x + \cdots$ and, dropping the higher powers, find

$$z_{vib} \sim T/\theta \quad (T \gg \theta). \tag{2.32}$$

The contribution of this one vibrational degree of freedom of this one molecule to the total free energy A is $-kT \ln z_{vib}$, so by the Gibbs–Helmholtz equation (1.14) its average contribution ε_{vib} to the system's energy is

$$\varepsilon_{vib} = -k \, d \ln z_{vib}/d(1/T). \tag{2.33}$$

From (2.31) and (2.32), and the definition of θ_{vib} in (2.30), we may see in particular what this contribution is at low and high temperatures. At low temperatures

2.3 Vibrational partition function

$\varepsilon_{vib} = \frac{1}{2}k\theta = \frac{1}{2}h\nu$, while at high temperatures $\varepsilon_{vib} = -k\,d\ln T/d(1/T) = k\,d\ln(1/T)/d(1/T) = k/(1/T) = kT$:

$$\varepsilon_{vib} \sim \begin{cases} \frac{1}{2}h\nu & (T \ll \theta) \\ kT & (T \gg \theta). \end{cases} \quad (2.34)$$

We recognize $\frac{1}{2}h\nu$ as the zero-point energy of a harmonic oscillator of frequency ν. At low temperatures, because of the Boltzmann factor $\exp(-\varepsilon/kT)$, almost all the oscillators are in their ground states, so the average contribution of each to the total energy will be just that ground-state energy $\frac{1}{2}h\nu$. (The word "zero" in the expression "zero-point energy" refers to the zero of *temperature*.) That accounts for the low-T limit in (2.34). At high temperatures, on the other hand, one is in the classical limit, where ε_{vib} no longer contains any reference to h. Instead, the average energy becomes just kT, the same for any harmonic oscillator, whatever its frequency (although how high T must be to count as high does depend on ν, because this limit requires $T \gg \theta_{vib} = h\nu/k$). Contrast this kT with the average energy $\frac{1}{2}kT$ per translational degree of freedom that we found in (2.19). The oscillator, unlike a free particle, is subject to a varying potential energy, and, on average, it has as much potential as kinetic energy. The average value of each in the classical limit is $\frac{1}{2}kT$, so the total $\varepsilon_{vib} = kT$ in this limit, as found in (2.34).

The energy ε_{vib} at any temperature, not just at the extremes of high and low T, may be obtained from the z_{vib} given by (2.29). It is shown plotted against T in Fig. 2.2. Note that at high and low T the plot is consistent with (2.34).

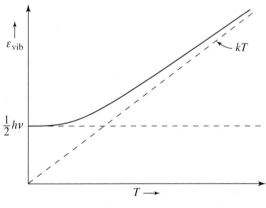

Fig. 2.2

30   2 The ideal gas

Exercise (2.3). Calculate the mean energy ε_{vib} for a single molecular vibrational degree of freedom of frequency ν, at temperature T. Plot $\varepsilon_{\text{vib}}/h\nu$ as a function of $kT/h\nu$ and compare with Fig. 2.2.

Solution. ε_{vib} is given by Eq. (2.33) with z_{vib} as given in Eq. (2.29). Then

$$\varepsilon_{\text{vib}}/h\nu = -\frac{d \ln z_{\text{vib}}}{d(h\nu/kT)} = \frac{d \ln z_{\text{vib}}^{-1}}{d(h\nu/kT)}$$

$$= \frac{d}{d(h\nu/kT)} \ln\left(e^{h\nu/2kT} - e^{-h\nu/2kT}\right)$$

$$= \frac{1}{2}\frac{e^{h\nu/2kT} + e^{-h\nu/2kT}}{e^{h\nu/2kT} - e^{-h\nu/2kT}} = \frac{1}{2}\frac{1 + e^{-h\nu/kT}}{1 - e^{-h\nu/kT}}.$$

This is plotted in the accompanying figure. It goes to $1/2$ as $T \to 0$ and is asymptotic to $kT/h\nu$ as $T \to \infty$. Except that the ordinate is here scaled by $h\nu$ and the abscissa by $h\nu/k$, this plot is the same as that in Fig. 2.2.

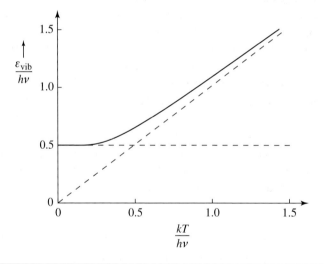

The average contribution c_{vib} of this one vibrational degree of freedom of this one molecule to the total heat capacity is

$$c_{\text{vib}} = d\varepsilon_{\text{vib}}/dT. \quad (2.35)$$

We can see qualitatively, from (2.35) and Fig. 2.2, that c_{vib} as a function of temperature is as shown in Fig. 2.3. The high-temperature limit is k, or R per

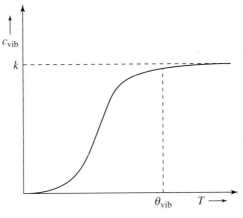

Fig. 2.3

mole. We shall wish to recall this later (Chapter 4) in connection with the law of Dulong and Petit.

Note that c_{vib} vanishes in the extreme quantum limit at low T. That is because $T \ll \theta_{vib} = h\nu/k$ means that the energy-level spacing $h\nu$ is much greater than the thermal energy kT. Excitation from the ground state is then a difficult and rare event; so when one measures the heat capacity by measuring the temperature rise that accompanies the absorption of heat, none of the added energy can go into that degree of freedom. Thus, at low T the vibration becomes "inert," or "frozen out," and ceases to contribute to the heat capacity (although it still has its zero-point energy). The fall-off in c_{vib} with decreasing temperature begins to be noticeable at about $T = \theta_{vib}$ and becomes rapidly more extreme as T decreases further (Fig. 2.3). Since for typical vibrations θ_{vib} is greater than 1000 K and may be several times that, it means that near 300 K the typical vibration contributes little to the total heat capacity.

2.4 Rotational partition function; *ortho*- and *para*-hydrogen

To the extent that a molecule's rotational and vibrational motions are independent, the partition function z_{int} associated with its internal (as distinct from translational) modes may be factorized as in (2.27) and the rotational degrees of freedom treated as those of a rigid rotor. Having already dealt with the vibrational factor in §2.3, we turn now to z_{rot}.

We consider first a diatomic molecule in the classical, high-temperature limit in which we may ignore the quantization of rotational energy levels. We know even from classical mechanics but also from quantum mechanics that the

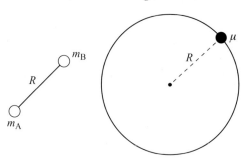

Fig. 2.4

rotation of a rigid diatomic whose individual atomic masses are m_A and m_B is dynamically equivalent to the motion of a single particle of reduced mass μ,

$$\mu = m_A m_B/(m_A + m_B), \tag{2.36}$$

confined to a sphere of radius R equal to the fixed internuclear distance in the molecule (Fig. 2.4). Then z_{rot} must be the same as the *translational* partition function of a particle of mass μ in two dimensions confined to a "box" of area equal to that of a sphere of radius R; i.e., to a two-dimensional box of area $4\pi R^2$. Then from (2.16) or (2.17) we obtain for the classical limit of z_{rot},

$$\begin{aligned} z_{\text{rot}} &= \left(\frac{2\pi \mu kT}{h^2}\right)^{2/2} \cdot 4\pi R^2 \\ &= 8\pi^2 IkT/h^2 \end{aligned} \tag{2.37}$$

with the moment of inertia I given by

$$I = \mu R^2. \tag{2.38}$$

(If r_A and r_B are the distances of the nuclei A and B from the center of mass of the molecule, (2.38) is equivalent to $I = m_A r_A^2 + m_B r_B^2$, as one may verify by explicit calculation from (2.36) and (2.38) together with $r_A + r_B = R$ and the center-of-mass condition $m_A r_A = m_B r_B$.)

The formula (2.37), however, is correct only for a heteronuclear, not for a homonuclear, diatomic. The reason is similar to that which led to the division by the product of factorials in (2.6). For a heteronuclear diatomic the dynamical state in which the molecule is rotating with a certain angular velocity and is, momentarily, in a particular orientation, is different from that in which it has the same angular velocity but an orientation different by 180° from that in the first state; while for a homonuclear diatomic those two states are identical, because

2.4 Rotational partition function

rotation by 180° merely exchanges the identical nuclei. Thus, while the whole of the area $4\pi R^2$ in (2.37) corresponds to distinct states of the heteronuclear diatomic, for the homonuclear molecule it counts every state twice, so the right-hand side of (2.37) should then be divided by 2. If we introduce a *symmetry number* σ,

$$\sigma = \begin{cases} 1, & \text{heteronuclear diatomic} \\ 2, & \text{homonuclear diatomic,} \end{cases} \quad (2.39)$$

then (2.37) should be replaced by

$$z_{\text{rot}} = \frac{8\pi^2 I k T}{\sigma h^2}. \quad (2.40)$$

We turn next to the evaluation of the fully quantum mechanical z_{rot} for a diatomic molecule, which we expect to reduce to (2.40) in the classical limit of high temperatures, or large moments of inertia (large masses and large sizes), or, formally, $h \to 0$.

The energy levels, indexed by the quantum number J associated with the total angular momentum, as given by quantum mechanics, are

$$\varepsilon_J = \frac{\hbar^2 J(J+1)}{2I} = \frac{h^2 J(J+1)}{8\pi^2 I} \quad (J = 0, 1, 2, \ldots), \quad (2.41)$$

with the moment of inertia I again given by (2.38) when the molecule is taken to be a rigid rotor. Each of these rotational energy levels is $(2J+1)$-fold degenerate, corresponding to the $2J+1$ possible values of the quantum number m_J associated with the component of the angular momentum in some arbitrarily chosen direction. The rotational energy depends only on the total angular momentum, not on its component in an arbitrary direction, hence only on J, not on m_J. Then for each J there are $2J+1$ identical terms in the rotational partition function $\Sigma \exp(-\varepsilon/kT)$. If these $2J+1$ terms are collected together, and the partition function is then written as a sum over J instead of as a sum over all distinct states, we then have for the diatomic molecule,

$$z_{\text{rot}} = \sum_{J=0}^{\infty} (2J+1) e^{-J(J+1)\theta/T}, \quad (2.42)$$

where θ, shorthand for θ_{rot} (to distinguish it from θ_{vib} in the preceding section), is defined by

$$\theta_{\text{rot}} = h^2/(8\pi^2 I k). \quad (2.43)$$

At low temperatures, $T \ll \theta_{\text{rot}}$, the sum in (2.42) is dominated by its terms of lowest J, each successive term being a successively smaller correction:

$$z_{\text{rot}} = 1 + 3e^{-2\theta/T} + 5e^{-6\theta/T} + \cdots . \qquad (2.44)$$

At high temperatures, $T \gg \theta_{\text{rot}}$, by the same kind of argument that allowed us to replace the translational partition-function sum by an integral (§2.2), we now obtain

$$\begin{aligned} z_{\text{rot}} &\sim \int_0^\infty (2J+1)e^{-J(J+1)\theta/T} \, dJ \\ &= \int_{J=0}^{J=\infty} e^{-J(J+1)\theta/T} \, dJ(J+1) \\ &= (T/\theta) \int_0^\infty e^{-x} \, dx \\ &= T/\theta \quad (T \gg \theta). \end{aligned} \qquad (2.45)$$

Recalling the definition of $\theta \, (= \theta_{\text{rot}})$ in (2.43), we see that this high-temperature limit of the quantum mechanical z_{rot} is the same as the classical z_{rot} in (2.37), as we expected it would be. But we know that (2.37) is right only for heteronuclear diatomics; for homonuclear ones, in the classical limit, the correct z_{rot} is given by (2.40) with $\sigma = 2$. Yet the fully quantum mechanical z_{rot} in (2.42) reduces to T/θ, not $T/2\theta$, in the high-T limit. Where might we find an additional factor of 1/2 in the high-temperature limit of the quantum mechanical z_{rot} for homonuclear diatomics?

The answer is that when the two nuclei are identical only half the states that are summed over in (2.42) – either those of even J or those of odd J – are allowed quantum states of the molecule. That is because the total wave function of the molecule, including its nuclear spin factors, must be either even or odd with respect to exchange of the identical nuclei: even, if the nuclear spin quantum number is integral (the nucleus contains an even number of nucleons: neutrons and protons), or odd if it is half-odd-integral (the nucleus consists of an odd number of nucleons); i.e., even, if the nuclei are bosons, or odd if they are fermions (cf. §2.1). Therefore, given the nature of the nuclei, and given the nuclear spin state, which specifies the relative orientation of the two nuclear spins and may itself be either symmetric or antisymmetric (even or odd) with respect to the exchange of the two nuclei, the rotational part of the total molecular eigenfunction must be definitely even or definitely odd with respect to rotation of the molecule by 180°, which also interchanges the nuclei. That means either even or odd with respect to the replacement of $\cos \chi$ by $-\cos \chi$, where χ is the angle the molecule makes with some arbitrarily chosen direction in space; and that, in turn, means that the molecule in any given nuclear-spin state may exist in rotational states of even J only or of odd J only.

For the diatom of ordinary hydrogen ($_1^1$H), where each nucleus is just a proton of spin quantum number 1/2, there are three possible symmetric states of the pair of spins (nuclear spin triplet) and only one possible antisymmetric nuclear spin state (spin singlet). Since protons are fermions, so that the total wave function must be odd in the exchange of the nuclei, it means that those H$_2$ molecules that find themselves in one of the three symmetric nuclear spin states can exist only in rotational states of odd J, while those that find themselves in the antisymmetric nuclear spin state can exist only in rotational states of even J. The former are named *ortho*-hydrogen and latter *para*-hydrogen. Thus, for any of the three kinds of *ortho*-H$_2$,

$$J = 1, 3, 5, \ldots \quad (\textit{ortho-}H_2), \qquad (2.46)$$

while for the single kind of *para*-H$_2$,

$$J = 0, 2, 4, \ldots \quad (\textit{para-}H_2). \qquad (2.47)$$

In either case, for H$_2$ or any other homonuclear diatomic, half the terms in (2.42) are missing; but whether it is the even or the odd J that are missing depends on which of the two possible species of that molecule, distinguished by the symmetry of the nuclear spin state, we are talking about. At low temperatures it makes a big difference which it is; the successive terms in the expansion (2.44) are rapidly decreasing, so it matters which ones are present and which are absent. Low temperature means $T \ll \theta_{\text{rot}}$, which for H$_2$ is 88 K. At high temperatures, by contrast, it does not matter which terms are missing from the sum (2.42) because the terms that contribute significantly to that sum vary slowly with J; that is what made it possible to replace the sum by an integral. That half the terms are nevertheless missing from (2.42) means that the partition function in this classical limit is only half as great as it is for a heteronuclear diatomic, for which all the terms are present. For the homonuclear diatomic, then, at high temperatures, (2.45) is to be replaced by $z_{\text{rot}} = T/2\theta$, exactly as in (2.40) with $\sigma = 2$.

Is the rule that any homonuclear diatomic molecule is limited to states of even J alone or odd J alone not violated in infra-red spectroscopy, where the selection rule is $\Delta J = \pm 1$? No, because a homonuclear diatomic, as one learns in spectroscopy, does not have an infra-red spectrum. But it does have a Raman spectrum, so is the rule not violated there? No, because there the selection rule is $\Delta J = 0, \pm 2$.

At high temperatures, which generally means $T \gg \theta_{\text{rot}}$ and so for H_2 means T well above 88 K, the three *ortho* nuclear-spin states and one *para* state are equally populated, so equilibrium hydrogen at high temperatures is a mixture of *ortho-* and *para-*hydrogen in the ratio 3:1. In pure hydrogen the magnetic interactions between molecules that could cause transitions between nuclear-spin states are so weak that such hydrogen may be cooled to arbitrarily low temperatures with the *ortho:para* ratio remaining 3:1. That is normal hydrogen. Paramagnetic impurities such as O_2, or a catalyst such as Pt on the surface of which H_2 molecules may dissociate, allowing the atoms to recombine with their nuclear spins in a different relative orientation, catalyze the *ortho–para* conversion. Cooled in the presence of such a paramagnetic impurity or such a catalyst, then, the hydrogen may have its *ortho:para* ratio continually adjust to that which would correspond to thermal equilibrium at the given low temperature. The successive terms in the low-temperature series expansion (2.44) for z_{rot} are the Boltzmann factors determining the relative populations of the energy levels $J = 0, 1, 2, \ldots$ at equilibrium, without yet taking account of the nuclear-spin degeneracy. The leading term, 1, comes from $J = 0$, which according to (2.47) is the lowest-lying rotational state of *para-*H_2, while the next term, $3\exp(-2\theta/T)$, comes from $J = 1$, which by (2.46) is the lowest-lying rotational state of *ortho-*H_2. There is then the additional factor of 3, from the nuclear-spin degeneracy, favoring the *ortho* species. Thus, at very low temperatures, where the further terms are negligible, the *ortho:para* ratio in equilibrium hydrogen is $9\exp(-2\theta/T)$, which is small when $T \ll \theta_{\text{rot}}$ and goes rapidly to 0 as $T \to 0$.

Cooled in such a way, then, hydrogen becomes pure *para-*H_2 at low temperatures, in contrast to "normal" hydrogen, which remains a mixture of *ortho* and *para* in the ratio 3:1. If the hydrogen is cooled in the presence of a catalyst that catalyzes the *ortho–para* conversion and thus allows it to become pure *para-*H_2 at low T, and if the catalyst is then removed and the hydrogen warmed back up to 298 K, say, the result is a sample of pure *para-*H_2 at normal laboratory temperatures. Pure *ortho-*H_2 cannot be prepared in this way.

Exercise (2.4). Calculate the rotational contributions to the heat capacities of normal hydrogen (3:1 mixture of *ortho* and *para*) and of equilibrium hydrogen at very low temperatures, $T \ll \theta_{\text{rot}}$.

Solution. $c_{\text{rot}} = d\varepsilon_{\text{rot}}/dT$ and $\varepsilon_{\text{rot}} = -k\, d\ln z_{\text{rot}}/d(1/T)$ (cf. (2.33) for ε_{vib}). At low temperatures, z_{rot} for *para-*H_2 consists of the even-J terms in the series (2.44), hence $z_{\text{rot},para} \sim 1 + 5\exp(-6\theta/T)$, while for *ortho-*$H_2$ it consists of the odd-J terms, $z_{\text{rot},ortho} \sim 3\exp(-2\theta/T) + 7\exp(-12\theta/T) = 3\exp(-2\theta/T) \times$

2.4 Rotational partition function

$[1 + (7/3)\exp(-10\theta/T)]$. The exponentials are all small, and $\ln(1+x) \sim x$ for small x. Thus, $\ln z_{\text{rot},para} \sim 5\exp(-6\theta/T)$ and $\ln z_{\text{rot},ortho} \sim \ln 3 - 2\theta/T + (7/3)\exp(-10\theta/T)$. Then carrying out the differentiations with respect to $1/T$ and T indicated above, we find

$$c_{\text{rot},para} = -k\, d\!\left(-30\,\theta e^{-6\theta/T}\right)\!/dT$$
$$= 180k(\theta/T)^2 e^{-6\theta/T}$$
$$c_{\text{rot},ortho} = -k\, d\!\left[-2\theta - (70/3)\theta e^{-10\theta/T}\right]\!/dT$$
$$= (700/3)k(\theta/T)^2 e^{-10\theta/T}.$$

At low temperatures $c_{\text{rot},para}$ is seen to be much greater than $c_{\text{rot},ortho}$ and it becomes infinitely greater as $T \to 0$ (although both go to 0). Then c_{rot} of *normal hydrogen* becomes entirely that of its *para* component as $T \to 0$; so with 1/4 of the molecules being *para*, we have, per mole, C_{rot} (normal H_2) $\sim 45R(\theta/T)^2 \exp(-6\theta/T)$.

For *equilibrium hydrogen* every molecule has access to every rotational state because its nuclear-spin state can adjust accordingly, with the result that every term in (2.42) is present in z_{rot}, but now each of the odd-J terms is to be multiplied by an additional factor of 3 to take account of the nuclear-spin degeneracy. The resulting partition function may be thought of as that for the rotational and nuclear-spin degrees of freedom together, although for simplicity we shall continue to call it z_{rot}, now $z_{\text{rot},equil}$. Then $z_{\text{rot},equil} \sim 1 + 9\exp(-2\theta/T)$ as $T \to 0$, so $\ln z_{\text{rot},equil} \sim 9\exp(-2\theta/T)$. Therefore

$$c_{\text{rot},equil} = -k\, d\!\left(-18\theta e^{-2\theta/T}\right)\!/dT$$
$$= 36k(\theta/T)^2 e^{-2\theta/T},$$

or C_{rot} (equilibrium H_2) $\sim 36R(\theta/T)^2 \exp(-2\theta/T)$ per mole of H_2. Although this, too, goes to 0 as $T \to 0$, it is nevertheless much greater than C_{rot} of normal H_2 at low temperatures.

All the heat capacities calculated in Exercise (2.4) are low-temperature approximations obtained by dropping all but the early terms in the low-temperature expansion (2.44) for z_{rot}. The higher-order terms would have provided corrections to the heat capacities that would have been infinitely smaller, as $T \to 0$, than the leading contributions, which are all that were explicitly calculated. Even these leading contributions, however ($45R(\theta/T)^2 \exp(-6\theta/T)$ for the rotational contribution to the molar heat capacity of normal H_2 and $36R(\theta/T)^2 \exp(-2\theta/T)$ for equilibrium H_2), go rapidly to 0 as $T \to 0$. That is because when $T \ll \theta_{\text{rot}}$ the energy-level spacing between the ground state and the first accessible excited state is much greater than kT. The rotational degrees of freedom then become "inert," or "frozen out," just as do vibrational degrees of freedom at low temperatures (§2.3), and cease contributing to the heat capacity.

For hydrogen, we noted, $\theta_{\text{rot}} = 88$ K. For any other molecule the moment of inertia I is so much greater than that for H_2 that its θ_{rot}, by (2.43), would be even much lower than 88 K. Thus, except for H_2 and perhaps D_2, for any molecule at practically any temperature of interest one is in the classical limit $T \gg \theta_{\text{rot}}$, with z_{rot} given simply by (2.40). That is even true of H_2 at most temperatures of practical interest; the *ortho–para* distinction is a rather specialized concern, important for the thermodynamic properties only of hydrogen, deuterium, and perhaps methane, which have exceptionally low moments of inertia, and then only at very low temperatures.

For the more practical case of $T \gg \theta_{\text{rot}}$, then, we may take $z_{\text{rot}} = 8\pi^2 IkT/(\sigma h^2)$ for diatomic molecules, and then calculate the contribution the rotations make to the energy and heat capacity per molecule, ε_{rot} and c_{rot}, at high temperatures:

$$\varepsilon_{\text{rot}} = -k \, \text{d} \ln z_{\text{rot}}/\text{d}(1/T) = kT \qquad (2.48)$$

$$c_{\text{rot}} = \text{d}\varepsilon_{\text{rot}}/\text{d}T = k, \qquad (2.49)$$

or RT and R per mole, respectively. These are independent of the symmetry number σ and are indeed also independent of the moment of inertia I: in this classical, high-temperature limit the rotational contributions to the energy and heat capacity are the same for every diatomic molecule. (That is not true of the entropy and free energy, which do depend on σ and I.) A diatomic molecule has two rotational degrees of freedom, so the contributions to ε_{rot} and c_{rot} per degree of freedom are $\frac{1}{2}kT$ and $\frac{1}{2}k$, the same as for each translational degree of freedom (§2.2). The rotations are free rotations, not subject to forces, just as the translations are free, but unlike vibrations (§2.3).

These results for diatomic molecules at high temperatures apply more generally to all linear molecules, even if, like CO_2 or acetylene, they are polyatomic, for these still have just two rotational degrees of freedom with a common moment of inertia I. (We shall presently discuss the question of why linear molecules, including diatomics, may be deemed to have only two rotational degrees of freedom. The atoms, after all, have some spatial extent, so is rotation about the molecule's internuclear line not a third rotational degree of freedom?) Now we turn to the evaluation of the rotational partition function of non-linear polyatomics, which have three (in general different) principal moments of inertia I_1, I_2, I_3 associated with rotation about each of three mutually perpendicular

2.4 Rotational partition function

principal axes. We shall limit ourselves to the high-temperature regime, which is almost always the only case of interest, and shall thus be content with the three-degree-of-freedom counterpart of the z_{rot} in (2.40).

Let us recall that the predecessor to the formula (2.40) was the first of the equations (2.37), which, with the definition of I in (2.38), could have been written as

$$z_{\text{rot}} \text{ (linear)} = \left(\sqrt{\frac{8\pi^2 I k T}{h^2}} \right)^2. \tag{2.50}$$

The two powers of the square root arose because the two rotations were equivalent to translation in two dimensions, while the two associated moments of inertia had the common value I. Now for three rotational degrees of freedom with three different moments of inertia I_1, I_2, and I_3 we might guess that the analogous z_{rot} (non-linear) would just be the product of three such square roots, one for each degree of freedom and each containing its own I_i ($i = 1, 2, 3$). That is *almost* right. A detailed calculation based on the mechanics of rotation of rigid three-dimensional bodies (which we shall omit) shows that that guess is right only up to an additional factor of $\sqrt{\pi}$; i.e.,

$$z_{\text{rot}}(\text{non-linear}) = \frac{\sqrt{\pi}(8\pi^2 k T)^{3/2} \sqrt{I_1 I_2 I_3}}{h^3 \sigma}, \tag{2.51}$$

where now a symmetry number σ, like that in (2.40), has been included.

Examples of non-linear molecules with symmetries requiring a symmetry number σ greater than 1 in (2.51) are H_2O and NH_3; see Fig. 2.5. For H_2O, rotation by $180°$ about the symmetry axis (the bisector of the HOH angle) produces a configuration identical with the starting configuration, so $\sigma = 2$. For NH_3, rotation by either $120°$ or $240°$ about the symmetry axis (an axis through the N, perpendicular to the plane of the three H's) produces a configuration identical with the starting one, so $\sigma = 3$.

With z_{rot} given by (2.51) we may now again calculate the rotational contributions to the energy and heat capacity as in (2.48) and (2.49). Thus, at ordinary

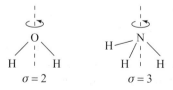

Fig. 2.5

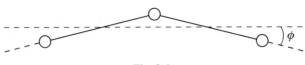

Fig. 2.6

temperatures, for non-linear molecules,

$$\varepsilon_{\text{rot}} = -k \, d \ln z_{\text{rot}}/d(1/T) = \frac{3}{2}kT \qquad (2.52)$$

$$c_{\text{rot}} = d\varepsilon_{\text{rot}}/dT = \frac{3}{2}k. \qquad (2.53)$$

There are now three degrees of rotational freedom, and the contributions to ε and c are again $\frac{1}{2}kT$ and $\frac{1}{2}k$ per degree of freedom.

Here is a paradox. Suppose a molecule is practically but not quite linear; imagine even an extreme (although entirely hypothetical) case in which a polyatomic molecule deviates from linearity by only a small fraction of one degree. This is pictured for a triatomic in Fig. 2.6, but we must imagine the angle ϕ as being only a fraction of a degree. Common sense tells us that the energy and heat capacity of a gas of such molecules would differ indiscernibly from those of a gas of the corresponding linear molecules, for which $\phi = 0$; yet if we are to believe (2.48), (2.49), (2.52), and (2.53), there would be a very substantial, easily measurable difference in the thermodynamic properties of those two gases, in particular, a difference of $\frac{1}{2}R$ in their molar heat capacities.

The resolution of the paradox is that if the angle ϕ pictured in Fig. 2.6 is very small, then rotation about one of the molecule's principal axes – the one represented schematically by the horizontal dashed line in the figure – would have associated with it a minute moment of inertia, hence widely spaced rotational energy levels. If the angle ϕ were small enough that spacing would greatly exceed kT, whereupon that rotational degree of freedom would be frozen out and would not contribute to the heat capacity. The formula (2.51) for z_{rot} (non-linear), which led to (2.52) and (2.53) for ε_{rot} and c_{rot}, assumes that we are at the classical, high-temperature limits for all three rotations. If one of them is in reality highly quantum mechanical because of a small moment of inertia, it must be treated quantum mechanically, not classically. That degree of freedom would then prove to be thermally inert, with the result that the $\frac{3}{2}kT$ and $\frac{3}{2}k$ of (2.52) and (2.53) would find themselves replaced by $\frac{2}{2}kT$ and $\frac{2}{2}k$. But these are just the values for linear molecules, in (2.48) and (2.49). Thus, if the angle ϕ in Fig. 2.6 were very small the properties of the molecule would indeed be nearly indistinguishable from those of a linear molecule, as dictated by common sense.

That same argument explains why, for a linear molecule, we may ignore the third rotational degree of freedom even though the atoms have some spatial extent. The internuclear line for a linear molecule is its third rotation axis, but electrons are so light, and nuclei are so small, that the moment of inertia associated with rotation about that axis would be minute and the corresponding energy-level spacing would be much greater than kT. Indeed, to excite the rotation of the electrons about that axis would require typical electronic excitation energies, with spacings of the order of some eV. But 1 eV/k is 11,600 K, which means that for that rotational degree of freedom not to be frozen out the temperature would have to be tens of thousands of degrees K. Excitation of the rotation of the nuclei about the internuclear line would be even much more extreme, requiring a nuclear excitation energy, which is typically in the MeV range.

Of course, when we say a molecule is linear we are referring only to a vibrational average; but the effective moment of inertia for rotation about the internuclear line that is imparted to the molecule by its bending vibrations is even then usually not great enough to require correcting (2.48) and (2.49). If that were not so, it would mean that we should not in the first place have assumed the rotations and vibrations to be independent of each other, as we did in (2.27).

2.5 The "law" of the equipartition of energies

We saw in §§2.2, 2.3, and 2.4 that the average contribution of each translational degree of freedom (of which there are three) to the energy and heat capacity is $\frac{1}{2}kT$ and $\frac{1}{2}k$ per molecule, respectively, or $\frac{1}{2}RT$ and $\frac{1}{2}R$ per mole; that at high temperatures ($T \gg \theta_{\text{vib}}$) the corresponding contributions by a vibrational degree of freedom are RT and R; and that at practically all temperatures ($T \gg \theta_{\text{rot}}$) those by a rotational degree of freedom are $\frac{1}{2}RT$ and $\frac{1}{2}R$. We saw also that the reason the results are the same for translation and rotation is that those energies are entirely kinetic, there being no forces that constrain those motions, while the vibrational contribution is twice as great because vibrational energy is equally kinetic and potential, on average, and the mean of each is $\frac{1}{2}kT$.

That ubiquitous $\frac{1}{2}kT$ in the classical, high-temperature limit arises from the "square terms" that each such mode of motion contributes to the total energy. The energy in any one translational degree of freedom is $\frac{1}{2}mv^2$, where v is the associated component of velocity and m is the molecule's mass. It is similarly $\frac{1}{2}I\omega^2$ for a rotational degree of freedom, where ω and I are the associated angular velocity and moment of inertia. In a vibrational degree of freedom, modeled as a harmonic oscillator, there are two such square terms, $\frac{1}{2}mv^2$ for the kinetic energy, where m is the mass of the oscillator and v its instantaneous velocity, and $\frac{1}{2}kx^2$ for the potential energy, where k is the force constant and x the displacement of the oscillator from its equilibrium position.

Each such square term is the square of a velocity or a displacement multiplied by some constant – a mass, moment of inertia, or force constant. We may represent any such term as ay^2, where y is then the velocity or the displacement and a the constant coefficient of y^2 in the expression for the energy. According to the Boltzmann distribution law the probability that the value of y (the velocity or displacement) will be found in the infinitesimal interval y to $y + dy$ when the temperature is T is proportional to $\exp(-\varepsilon/kT)dy$, where the energy ε associated with y is now ay^2 and where k here is again Boltzmann's constant (not to be confused with the force constant k). Then the mean value of ay^2, that is, the average contribution $\bar{\varepsilon}$ of that one part of the molecule's energy to the total energy of the system, is

$$\bar{\varepsilon} = \frac{\int_{-\infty}^{\infty} ay^2 \exp(-ay^2/kT)\, dy}{\int_{-\infty}^{\infty} \exp(-ay^2/kT)\, dy}$$

$$= kT \int_{-\infty}^{\infty} x^2 e^{-x^2} dx \bigg/ \int_{-\infty}^{\infty} e^{-x^2} dx$$

$$= \frac{1}{2}kT. \qquad (2.54)$$

(That the integral in the numerator is half that in the denominator may be seen on integrating the former by parts.) We thus see that whatever the character of y – whether it be a velocity or a displacement – and whatever the constant coefficient a – whether it be a mass, a moment of inertia, or a force constant – as long as the term ay^2 is a separate contribution to the total energy and the temperature is high enough for that degree of freedom to be treated classically, such a square term contributes $\frac{1}{2}kT$ to the average energy.

This is the "law" of the equipartition of energy, with "law" in quotation marks because of all the conditions required for its validity. In particular, for vibrations or rotations it requires that $T \gg \theta_{\text{vib}}$ or $T \gg \theta_{\text{rot}}$, the second of which usually holds but the first of which rarely does, as we know (§2.3); and it requires in addition that a vibration be harmonic, or else the associated potential energy would not be simply quadratic in the displacement.

Exercise (2.5). The ring-puckering vibration in a molecule with a four-membered ring is close to being dynamically independent of the other vibrations of the molecule. In cyclobutanone and in trimethylene oxide these ring-puckering vibrations are almost pure quartic oscillators; i.e., the potential

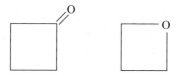

energy of puckering is almost exactly proportional to the fourth power of a coordinate that measures the extent of pucker. What is the contribution to the heat capacity made by such a vibration at high temperatures?

Solution. With x the ring-puckering coordinate, the potential energy is of the form αx^4 with some constant α. Then the mean potential energy in this degree of freedom, in the classical, high-temperature limit, is

$$\bar{\varepsilon}_{\text{pot}} = \frac{\int_{-\infty}^{\infty} \alpha x^4 \exp(-\alpha x^4/kT)\,dx}{\int_{-\infty}^{\infty} \exp(-\alpha x^4/kT)\,dx}$$

$$= kT \int_{-\infty}^{\infty} y^4 e^{-y^4}\,dy \bigg/ \int_{-\infty}^{\infty} e^{-y^4}\,dy.$$

But

$$\int_{-\infty}^{\infty} y^4 e^{-y^4}\,dy = \int_{y=-\infty}^{y=\infty} -\frac{1}{4} y\, d e^{-y^4}$$

$$= \frac{1}{4} \int_{-\infty}^{\infty} e^{-y^4}\,dy \quad \text{(integration by parts)},$$

so $\bar{\varepsilon}_{\text{pot}} = \frac{1}{4}kT$. In the meantime the kinetic energy is still $\frac{1}{2}m(dx/dt)^2 = \frac{1}{2}mv^2$, so the mean kinetic energy is $\bar{\varepsilon}_{\text{kin}} = \frac{1}{2}kT$. The mean total energy in that vibration is then $\bar{\varepsilon} = \frac{3}{4}kT$. The contribution of that vibration to the high-temperature heat capacity is then $\frac{3}{4}k$ per molecule or $\frac{3}{4}R$ per mole.

2.6 Partition function with excited electronic states

It was assumed in §2.3 [Eq. (2.27)] that only the electronic ground state was significantly populated, so that the only internal (as distinct from translational) modes that needed to be considered were vibrational and rotational. Even when that holds we might still wish to, or need to, take account of a possible degeneracy of the ground electronic state. If its degeneracy is g_0 it means that every term we have so far included in the partition function $z_{\text{int}} = \Sigma \exp(-\varepsilon_{\text{int}}/kT)$

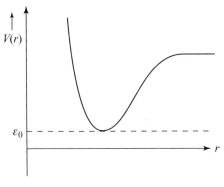

Fig. 2.7

should really have been counted g_0 times; i.e., our z_{int} should really have been taken to be

$$z_{int} \simeq g_0 z_{vib} z_{rot}, \tag{2.55}$$

with z_{vib} and z_{rot} still those calculated in §§2.3 and 2.4.

With $\bar{\varepsilon}_{int} = -k\, d \ln z_{int}/d(1/T)$, we see that the extra constant factor g_0 in (2.55) does not contribute to the energy, and so also not to the heat capacity. It does contribute to the entropy and free energy. If a_{int} and s_{int} are the corresponding contributions, per molecule, to A and S, then $a_{int} = -kT \ln z_{int}$ and $s_{int} = (\bar{\varepsilon}_{int} - a_{int})/T$, so the extra factor g_0 in (2.55) contributes the extra additive term $-kT \ln g_0$ to the free energy per molecule and $k \ln g_0$ to the entropy per molecule.

In Fig. 2.7 there is shown the potential curve for the ground electronic state of a stable diatomic molecule: potential energy $V(r)$ as a function of internuclear distance r. This figure may be taken as symbolic, also, for the many-dimensional potential-energy surface of a polyatomic molecule. Its minimum is shown as at an energy ε_0. This ε_0 may be assigned any value; there is always an arbitrary additive constant in any potential energy. The vibrational and rotational partition functions calculated in §§2.3 and 2.4 were calculated with the energies ε_{int} on a scale on which the energy ε_0 at the bottom of the potential well in Fig. 2.7 is 0. It is on such a scale, for example, that the low-lying vibrational energies, approximated as those of a harmonic oscillator, are $\varepsilon_v = (v + \frac{1}{2})h\nu$ ($v = 0, 1, 2, \ldots$).

If for any reason we should wish to assign to the energy at the bottom of the well in Fig. 2.7 some value ε_0 other than 0, then every ε_{int} in the partition function sum $z_{int} = \Sigma \exp(-\varepsilon_{int}/kT)$ would be greater by ε_0 than it was taken to be in the calculations in §§2.3 and 2.4, with the result that z_{int} would differ

2.6 Partition function with excited electronic states

by a factor $\exp(-\varepsilon_0/kT)$ from that calculated earlier. If z_{vib} and z_{rot} are those calculated earlier, then (2.55) must now read

$$z_{\text{int}} \simeq g_0 e^{-\varepsilon_0/kT} z_{\text{vib}} z_{\text{rot}}. \tag{2.56}$$

The effect is to increment $\bar{\varepsilon}_{\text{int}}$ by $-k\,\mathrm{d}\ln[\exp(-\varepsilon_0/kT)]/\mathrm{d}(1/T) = \varepsilon_0$, as expected, and to increment a_{int} also by $-kT\ln\exp(-\varepsilon_0/kT) = \varepsilon_0$. It leaves $s_{\text{int}}[=(\bar{\varepsilon}_{\text{int}} - a_{\text{int}})/T]$ and the heat capacity unaffected.

If we were to continue to consider the ground electronic state alone, by restricting consideration to temperatures so low that excited electronic states could be ignored, the foregoing changes introduced by including the electronic degeneracy g_0 would be trivial and those associated with the choice of an arbitrary ε_0 would have no physical consequences at all. As soon as excited electronic states enter the picture, though, we must take careful account of their degeneracies and energies *relative* to those of the ground state. Thus, if z_{int_0}, $z_{\text{int}_1}, \ldots$ are internal partition functions calculated in the individual electronic states $0, 1, \ldots$, each with its own private zero of energy, if these zeros of energy, on a common scale, are at $\varepsilon_0, \varepsilon_1, \ldots$, and if the electronic-state degeneracies are $g_0, g_1, \ldots$, then the full internal partition function of the molecule is

$$z_{\text{int}} = g_0 e^{-\varepsilon_0/kT} z_{\text{int}_0} + g_1 e^{-\varepsilon_1/kT} z_{\text{int}_1} + \cdots. \tag{2.57}$$

The simplest form of (2.57) is that for a monatomic gas, for which the only internal states are electronic. Then if on some scale the atomic energy levels are $\varepsilon_0, \varepsilon_1, \ldots$, and the corresponding degeneracies are $g_0, g_1, \ldots$, the internal partition function for the atom is

$$z_{\text{int}} = g_0 e^{-\varepsilon_0/kT} + g_1 e^{-\varepsilon_1/kT} + \cdots. \tag{2.58}$$

Multiplied by the z_{trans} given by (2.17), this would then, by (2.9), be the whole single-molecule partition function z for this monatomic gas. The factor z_{int} as given by (2.58) is fully quantum mechanical while the factor z_{trans} as given by (2.17) is wholly classical. The zero on the energy scale for $\varepsilon_0, \varepsilon_1, \ldots$ is arbitrary but the differences $\varepsilon_1 - \varepsilon_0$, etc., are all definite, determinable energies.

Exercise (2.6). The two lines of the "D-line" doublet in the emission spectrum of sodium are at 5895.93 and 5889.96 Å. They arise from transitions to the ground state from the closely spaced (spin-orbit-split) $^2P_{3/2}$ and $^2P_{1/2}$ levels, of which the former lies higher. Those are the first two excited-state energy levels of the sodium atom. At what temperature in sodium vapor would those two levels be equally populated?

Solution. The terms in the partition function (2.58) are the Boltzmann factors giving the relative populations of the levels 0, 1, 2, ... If the $^2P_{1/2}$ and $^2P_{3/2}$ levels are numbered 1 and 2, respectively, and if their energies are ε_1 and ε_2 and their degeneracies are g_1 and g_2, then their populations will be equal when $\exp[-(\varepsilon_2 - \varepsilon_1)/kT] = g_1/g_2$. The degeneracies are $g_1 = 2 \cdot (1/2) + 1 = 2$ and $g_2 = 2 \cdot (3/2) + 1 = 4$. The energy difference $\varepsilon_2 - \varepsilon_1 = \varepsilon_2 - \varepsilon_0 - (\varepsilon_1 - \varepsilon_0)$ with ε_0 the energy of the ground state; so $\varepsilon_2 - \varepsilon_1 = hc[(5889.96 \text{ Å})^{-1} - (5895.93 \text{ Å})^{-1}]$. The temperature in question is then

$$\begin{aligned} T &= (\varepsilon_2 - \varepsilon_1)/[k \ln(g_2/g_1)] \\ &= hc[(5889.96 \text{ Å})^{-1} - (5895.93 \text{ Å})^{-1}]/(k \ln 2) \\ &= 36 \text{ K}. \end{aligned}$$

Only at such an unreasonably low temperature could the minute energy difference that would favor the lower level of the ^2P doublet overcome the bias, due to its greater degeneracy, that favors the upper one. At any reasonable temperature $\varepsilon_2 - \varepsilon_1$ is so much less than kT that the population of the $^2P_{3/2}$ level is just twice as great as that of the $^2P_{1/2}$ level.

3
Chemical equilibrium in ideal-gas mixtures

3.1 Thermodynamic preliminaries; the equilibrium constant

One of the early applications of statistical mechanics to chemistry was in the calculation of the equilibrium constants of gas-phase reactions from spectroscopic data. This was a dominant theme in the statistical mechanics of the 1930s, the decade after the invention of quantum mechanics. Quantum mechanics had provided the theoretical framework within which the mechanical properties of molecules could be inferred from their optical and microwave spectra. The formulas for the partition functions of ideal gases derived in Chapter 2 will be the elements from which the equilibrium constants will be constructed.

First, we must remind ourselves of the status and meaning of an equilibrium constant in classical thermodynamics. A general chemical equilibrium may be represented by

$$\nu_a A + \nu_b B + \cdots \rightleftharpoons \nu_x X + \nu_y Y + \cdots \qquad (3.1)$$

where $A, B, \ldots, X, Y, \ldots$ stand for the chemical formulas of the reacting species while $\nu_a, \nu_b, \ldots, \nu_x, \nu_y, \ldots$ are their respective coefficients in the balanced chemical equation (stoichiometric coefficients). At equilibrium

$$\nu_a \mu_a + \nu_b \mu_b + \cdots = \nu_x \mu_x + \nu_y \mu_y + \cdots \qquad (3.2)$$

where μ_a is the chemical potential of the species A in the equilibrium mixture, μ_b that of B, etc. Equation (3.2) is the condition for chemical equilibrium in the reaction (3.1), and holds generally whatever the conditions of the reaction: no matter whether it be in a dilute gas, in a highly compressed, non-ideal gas, in a dilute or concentrated liquid solution, etc., and whether it be homogeneous (that is, in one phase) or heterogeneous.

Now to specialize, we consider only reactions among dilute species $A, B, \ldots, X, Y, \ldots$; i.e., among components of a dilute gas or of an ideal-dilute solution, for which, as one learns in thermodynamics, the chemical potentials are logarithmic functions of the concentrations,

$$\mu_a = kT \ln(c_a/c_a^\circ), \text{ etc.} \tag{3.3}$$

Here c_a is the concentration of the species A in the reaction mixture; for convenience, and in anticipation of the statistical mechanical treatment in the next section, we take it to be the number density, i.e., the number of A molecules, N_A, per unit volume V:

$$c_a = N_A/V, \text{ etc.} \tag{3.4}$$

The quantity c_a° in (3.3), which also has the dimensions of a concentration, is independent of the concentration of A or of the concentration of any other of the dilute species in the reaction mixture; it is some function of the temperature alone, characteristic of the species A and of its interaction with the solvent (if any). Note that the prefactor in (3.3) is kT, not RT, so that these are molecular, not molar, chemical potentials, a choice that we again make for convenience in anticipation of the statistical mechanical treatment later.

A crucial point here is that, while the form (3.3) for the chemical potential of a dilute species is known to thermodynamics, which recognizes that the quantity c_a° is some function of temperature alone, independent of the concentrations, classical thermodynamics alone does not know what function of the temperature c_a° is. Thus, it knows the concentration dependence, but cannot evaluate the temperature dependence, of the chemical potential. It is statistical thermodynamics, as we shall see later (§3.2), that, at least for an ideal gas, explicitly calculates those functions of temperature, c_a°, c_b°, etc. It does so in terms of molecular parameters obtainable from spectroscopy.

In the meantime, the equilibrium condition (3.2), with the special form (3.3) for the chemical potentials of the dilute species, becomes

$$\nu_a kT \ln(c_a/c_a^\circ) + \cdots = \nu_x kT \ln(c_x/c_x^\circ) + \cdots. \tag{3.5}$$

Divide this through by the common factor kT, rewrite each $\nu \ln(c/c^\circ)$ as $\ln(c/c^\circ)^\nu$, and take the exponential of both sides, thus obtaining

$$(c_a/c_a^\circ)^{\nu_a}(c_b/c_b^\circ)^{\nu_b} \cdots = (c_x/c_x^\circ)^{\nu_x}(c_y/c_y^\circ)^{\nu_y} \cdots, \tag{3.6}$$

or

$$\frac{c_x^{\nu_x} c_y^{\nu_y} \cdots}{c_a^{\nu_a} c_b^{\nu_b} \cdots} = \frac{c_x^{\circ \nu_x} c_y^{\circ \nu_y} \cdots}{c_a^{\circ \nu_a} c_b^{\circ \nu_b} \cdots}. \tag{3.7}$$

The right-hand side of (3.7) consists of factors that are functions of temperature and are characteristic of the individual species in the chemical equilibrium (and of their interactions with the solvent, if there is one), but are themselves independent of the concentrations $c_a, c_b, \ldots$. The condition (3.7) on the equilibrium concentrations $c_a, c_b, \ldots$ is the law of mass action. The function of temperature,

$$K_c = \frac{c_x^{\circ\,\nu_x} c_y^{\circ\,\nu_y} \cdots}{c_a^{\circ\,\nu_a} c_b^{\circ\,\nu_b} \cdots}, \qquad (3.8)$$

is the equilibrium constant. It is not itself calculable by classical thermodynamics because the individual c°'s, as remarked above, are not.

If the reaction occurs in the gas phase the concentrations c_a, etc., are related to the respective partial pressures p_a, etc., by $p_a = c_a kT$. Then the equilibrium condition (3.7) may equally well be written

$$\frac{p_x^{\nu_x} p_y^{\nu_y} \cdots}{p_a^{\nu_a} p_b^{\nu_b} \cdots} = (kT)^{\nu_x + \nu_y + \cdots - (\nu_a + \nu_b + \cdots)} K_c = K_p, \qquad (3.9)$$

thus defining the alternative equilibrium constant K_p. When the sum $\nu_a + \nu_b + \cdots$ of the stoichiometric coefficients on the left-hand side of the balanced equation (3.1) equals the sum $\nu_x + \nu_y + \cdots$ of those on the right, there is no difference between K_p and K_c, as we see from the second equality in (3.9). Then also that equilibrium constant is dimensionless, for then the net power of concentration or partial pressure in the numerator of the mass-action expression is the same as that in the denominator.

Our task now is to evaluate the c°'s for the case in which the chemical equilibrium occurs in the gas phase. That is done in the following section.

3.2 Equilibrium constants from partition functions

We saw in Chapter 2, Eq. (2.6), that for an ideal gas, in the almost universally applicable approximation of the "Boltzmann statistics," the total partition function Z is expressible in terms of the single-molecule partition functions $z_1, z_2, \cdots$ by

$$Z = \frac{1}{N_1! N_2! \cdots} z_1^{N_1} z_2^{N_2} \cdots, \qquad (3.10)$$

where the gas mixture consists of N_i molecules of species i. From the formula $A = -kT \ln Z$ for the Helmholtz free energy [Eq. (1.16) of Chapter 1], and the formulas (1.20) for the chemical potentials μ_i, we obtain

$$\mu_i = -kT \left[\partial \ln \left(\frac{1}{N_i!} z_i^{N_i} \right) \Big/ \partial N_i \right]_{T,V}. \qquad (3.11)$$

As usual [cf. Chapter 2, Eqs. (2.21)–(2.24)], we may approximate $\ln(N_i!)$ by the asymptotic formula

$$\ln(N_i!) \sim N_i \ln N_i - N_i, \tag{3.12}$$

since N_i is very large. Then carrying out the indicated differentiation in (3.11),

$$\mu_i = -kT(-\ln N_i + \ln z_i) = kT \ln(N_i/z_i). \tag{3.13}$$

This is the chemical potential of the species i in the ideal-gas mixture when there are N_i molecules of that species in the mixture; z_i is the single-molecule partition function of species i. It is characteristic of the ideal gas that μ_i depends on the number and nature of the molecules of species i only.

If we now compare (3.13) with the formula (3.3) from classical thermodynamics, with the concentrations defined by (3.4), we arrive at the statistical mechanical evaluation of the quantities $c_a^\circ, c_b^\circ, \ldots$:

$$c_a^\circ = z_a/V, \text{ etc.} \tag{3.14}$$

That is, the quantity c_a°, which was some function of the temperature alone characteristic of the species A, and had the dimensions of a concentration (number per unit volume), has turned out to be the single-molecule partition function of molecules of that species, per unit volume. How do we know that z_a/V is a function of temperature alone? In particular, why is there no volume dependence in it? That is because, by Eq. (2.9) of the preceding chapter, each z is a product $z_{\text{trans}} z_{\text{int}}$ of translational and internal partition functions, and each z_{trans}, by (2.17), is V times a function of T, while each z_{int} is a function of T alone.

(As an aside, we may here remark that we now see why the function ϕ in Chapter 1, Eq. (1.15), must be essentially independent of $N_1, N_2, \ldots$, just as we saw in §2.2, from the known ideal-gas equation of state, that it must be independent of V. If ϕ depended other than linearly on $N_1, N_2, \ldots$ there would be an extra composition-dependent term in the chemical potential μ_i in (3.13) and μ_i would no longer have the form (3.3) required by classical thermodynamics. If ϕ depended only linearly on $N_1, N_2, \ldots$ the extra term would depend only on T, to which it would be proportional, and so could be absorbed in the c°; and, indeed, ϕ can be chosen to be an arbitrary linear function of $N_1, N_2, \ldots$ – but a universal function, the same for all substances and all mixtures – without physical consequence. Such a choice for ϕ other than the conventional $\phi = 0$ is connected with the arbitrary zero of entropy; cf. §1.3.)

With (3.14), the formula (3.8) for K_c now becomes

$$K_c = \frac{(z_x/V)^{\nu_x}(z_y/V)^{\nu_y}\cdots}{(z_a/V)^{\nu_a}(z_b/V)^{\nu_b}\cdots}, \tag{3.15}$$

3.2 Equilibrium constants from partition functions

with a corresponding formula for K_p from the second of the equalities (3.9). We may state the prescription (3.15) in words: To find the equilibrium constant K_c for the dilute-gas-phase reaction $\nu_a A + \cdots \rightleftharpoons \nu_x X + \cdots$, evaluate the partition functions z_a, etc., of the various species by the formulas of Chapter 2; cancel the factors V from the translational factors of each of those partition functions; raise the remaining factors (now functions of the temperature alone) to powers equal to the stoichiometric coefficients of the respective species in the balanced chemical equation; and then form the indicated ratio of products of those powers.

In §2.6, where we considered the possibility of electronic excitation, we saw how important it is to refer the energies of all states to a common energy scale with only a single arbitrary zero. Now that we are dealing with a chemical reaction it is likewise of the greatest importance that in evaluating the partition functions $z_a, \ldots, z_x, \ldots$ the energy levels of the reactant and product species be referred to a common energy scale.

Suppose, as an example, that the reaction interconverts four diatomic species, $AB + CD \rightleftharpoons AC + BD$, and that, referred to a common energy scale, the minima in the potential curves of the four are at $\varepsilon_1, \varepsilon_2, \varepsilon_3$, and ε_4, respectively (cf. Fig. 2.7). Then if z_{AB}, z_{CD}, etc., are the partition functions of AB, CD, etc., each calculated with the minimum of its potential curve taken to be energy 0 – as is the case in the formulas for the vibrational and rotational factors in z derived in §§2.3 and 2.4 – the correct partition functions for use in calculating K_c from (3.15) would be $z_{AB} \exp(-\varepsilon_1/kT)$, $z_{CD} \exp(-\varepsilon_2/kT)$, etc. [cf. Eq. (2.56)]. The result for the equilibrium constant K (this is a case in which K_c and K_p are a common, dimensionless K) would then be

$$K = \frac{(z_{AC}/V)(z_{BD}/V)}{(z_{AB}/V)(z_{CD}/V)} e^{-(\varepsilon_3+\varepsilon_4-\varepsilon_1-\varepsilon_2)/kT}. \qquad (3.16)$$

The energy difference $\Delta\varepsilon = \varepsilon_3 + \varepsilon_4 - \varepsilon_1 - \varepsilon_2$ in the exponent is (except for zero-point-energy corrections) very nearly the same as the reaction energy, ΔU, or the enthalpy of reaction, ΔH. The factor $\exp(-\Delta U/kT)$ or $\exp(-\Delta H/kT)$ is normally the most important factor in determining the overall magnitude of an equilibrium constant. Thus, it would be folly to evaluate the partition functions in (3.15) with meticulous care while neglecting to use a common energy scale for reactants and products or to append the appropriate factor $\exp(-\Delta\varepsilon/kT)$.

Exercise (3.1). Show that the equilibrium constant K_c for the gas-phase dissociation of a diatomic molecule, $A_2 \rightleftharpoons 2A$, at temperatures T much higher than the molecule's θ_{rot} but much lower than its θ_{vib} and too low for significant electronic excitation of either the molecule or the atom, is

$$K_c = \frac{g_A^2}{g_{A_2}} \frac{\sqrt{mkT/\pi}}{2hR^2} e^{-D/kT},$$

where D is the energy of dissociation of the molecule from its ground vibrational state, m is the mass of the atom, R is the equilibrium internuclear distance in the molecule, and g_A and g_{A_2} are the degeneracies of the ground electronic states of the atom and molecule.

Solution. The accompanying figure shows the molecule's ground-state potential energy $V(r)$ as a function of internuclear distance r. On the scale of energy on which the bottom of the vibrational potential of A_2 is at 0, the electronic energy of 2A is $D + \frac{1}{2}h\nu$ (see figure), so that of A is $\frac{1}{2}(D + \frac{1}{2}h\nu)$. Then from Eqs. (2.17), (2.31), (2.40), and (2.56),

$$z_A = \underbrace{\left(\frac{2\pi mkT}{h^2}\right)^{3/2} V}_{\text{trans}} \cdot \underbrace{g_A e^{-\frac{1}{2}(D+\frac{1}{2}h\nu)/kT}}_{\text{electronic}}$$

$$z_{A_2} = \underbrace{\left[\frac{2\pi(2m)kT}{h^2}\right]^{3/2} V}_{\text{trans}} \cdot \underbrace{\frac{8\pi^2 IkT}{h^2\sigma}}_{\text{rot}(T \gg \theta_{\text{rot}})} \cdot \underbrace{e^{-\frac{1}{2}h\nu/kT}}_{\text{vib}(T \ll \theta_{\text{vib}})} \cdot \underbrace{g_{A_2}}_{\text{electronic}},$$

where

$$I = \mu R^2 = \frac{m \cdot m}{m+m} R^2 = \frac{1}{2}mR^2$$

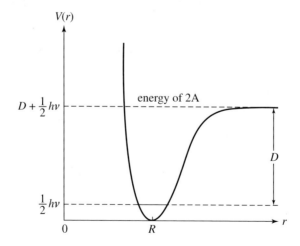

3.2 Equilibrium constants from partition functions

and $\sigma = 2$. Then

$$K_c = \frac{(z_A/V)^2}{z_{A_2}/V}$$

$$= \frac{g_A^2}{g_{A_2}} \left(\frac{\pi mkT}{h^2}\right)^{3/2} \frac{h^2}{2\pi^2 m R^2 kT} e^{-D/kT}$$

$$= \frac{g_A^2}{g_{A_2}} \frac{\sqrt{mkT/\pi}}{2hR^2} e^{-D/kT}.$$

Exercise (3.2). Calculate the equilibrium constant K_p for the dissociation $I_2 \rightleftharpoons 2I$ at 800 K. The ground state of the iodine atom is $^2P_{3/2}$, hence 4-fold degenerate, while that of the molecule is $^1\Sigma_g^+$, hence non-degenerate. (The notation here is as used in spectroscopy.) The excited electronic states of I and I_2 may be neglected at 800 K. For the molecule, $\theta_{\text{rot}} = 0.054$ K, $\theta_{\text{vib}} = 308$ K, and the dissociation energy from the ground vibrational state is $D = 1.5417$ eV (1 eV $= 1.6022 \times 10^{-19}$ J).

Solution. This is a variation of the calculation in the solution to Exercise (3.1). Here $g_I = 4$ and $g_{I_2} = 1$, while θ_{vib} is now not much greater – indeed, it is less – than T, so the full vibrational partition function as given in Eq. (2.29) must be used,

$$z_{\text{vib}} = \left(e^{hv/2kT} - e^{-hv/2kT}\right)^{-1},$$

and not approximated by $e^{-hv/2kT}$. Recalling that $\theta_{\text{vib}} = hv/k$ and $\theta_{\text{rot}} = h^2/(8\pi^2 Ik)$ [Eqs. (2.30) and (2.43)], with $I = \mu R^2 = \frac{1}{2}mR^2$ [$m =$ mass of the atom; Exercise (3.1)], the results for z_A and z_{A_2} in Exercise (3.1) now become

$$z_I = \left(\frac{2\pi mkT}{h^2}\right)^{3/2} V \cdot 4e^{-\frac{1}{2}(D/kT + \theta_{\text{vib}}/2T)}$$

$$z_{I_2} = \left[\frac{2\pi(2m)kT}{h^2}\right]^{3/2} V \cdot \frac{T}{2\theta_{\text{rot}}} \cdot \left(e^{\theta_{\text{vib}}/2T} - e^{-\theta_{\text{vib}}/2T}\right)^{-1} \cdot 1,$$

so that

$$K_c = \frac{(z_I/V)^2}{z_{I_2}/V} = 32\left(\frac{\pi mkT}{h^2}\right)^{3/2} \frac{\theta_{\text{rot}}}{T}\left(1 - e^{-\theta_{\text{vib}}/T}\right)e^{-D/kT}.$$

But for this reaction, $I_2 \rightleftharpoons 2I$, we have

$$K_p = kT K_c,$$

according to Eqs. (3.1) and (3.9), so

$$K_p = 32\left(\frac{\pi mkT}{h^2}\right)^{3/2} k\theta_{\text{rot}}\left(1 - e^{-\theta_{\text{vib}}/T}\right)e^{-D/kT}.$$

The mass m of the iodine atom (atomic weight 127 g/mol) is

$$m = \frac{0.127}{6.022 \times 10^{23}} \text{ kg} = 2.11 \times 10^{-25} \text{ kg}.$$

Then with $T = 800$ K, $\theta_{\text{rot}} = 0.054$ K, $\theta_{\text{vib}} = 308$ K, $D = 1.5417 \times 1.6022 \times 10^{-19}$ J $= 2.470 \times 10^{-19}$ J, $k = 1.3807 \times 10^{-23}$ J/K, and $h = 6.6261 \times 10^{-34}$ J s, we calculate

$$K_p = 3.19 \text{ Pa} = 3.15 \times 10^{-5} \text{ atm}.$$

(1 atm $= 1.01325 \times 10^5$ Pa.)

4

Ideal harmonic solid and black-body radiation

4.1 Ideal harmonic crystal

We may usefully think of a crystalline solid as a single huge molecule in which the motions of the atoms are decomposable into nearly independent normal modes of vibration. Weak interactions between the modes allow energy to flow between them, leading to thermal equilibrium among them, just as the weak interactions and rare collisions between molecules in an otherwise ideal gas are necessary in order to establish the equilibrium properties of the gas including the Maxwell velocity distribution. Just as there is no further reference to those interactions in the equilibrium properties of the ideal gas, so will there be no further reference to the weak, anharmonic interactions between the otherwise independent harmonic oscillators that constitute this ideal crystal.

If the crystal consists of N atoms there will be $3N - 6$ such normal modes of vibration, but since N is very large we may ignore the 6, which is associated with the translation and rotation of the crystal as a whole, and say that there will be $3N$ vibrational modes. Precisely because N is large, it is pointless to try to enumerate the modes and find the frequency ν_i ($i = 1, \ldots, 3N$) of each one. Instead, we recognize that there is virtually a continuum of frequencies ν, and we characterize the crystal by a frequency-distribution function $g(\nu)$ such that $g(\nu)d\nu$ is the fraction of all the $3N$ modes that have frequencies in the infinitesimal frequency interval ν to $\nu + d\nu$. Given the crystal structure, the force constants, and the masses of the atoms, there is in the asymptotic limit $N \to \infty$ a well-defined frequency distribution $g(\nu)$ as a function of the continuous variable ν.

A typical example of such a frequency-distribution function $g(\nu)$ is that for LiF, shown here in Fig. 4.1. It was calculated from a theoretical model of the crystal in which the parameters were chosen to make the dispersion curves for inelastic slow-neutron scattering implied by the model fit those that were

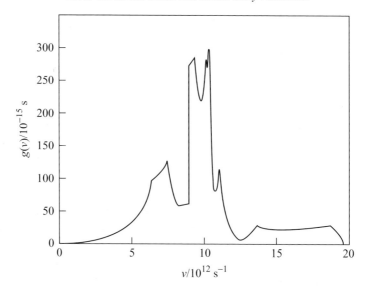

Fig. 4.1 Adapted from G. Dolling *et al.*, loc. cit. (1968). Reproduced with permission.

measured experimentally at 298 K. [G. Dolling, H.G. Smith, R.M. Nicklow, P.R. Vijayaraghavan, and M.K. Wilkinson, *Phys. Rev.* **168** (1968) 970.] Since $g(\nu)d\nu$ is dimensionless $g(\nu)$ itself has the dimensions of time; the $g(\nu)$ scale in the figure is in units of 10^{-15} s (femtoseconds), as indicated, while the ν scale is in units of 10^{12} s^{-1} (10^{12} Hz).

Such curves, as we see, may have a complicated structure, with kinks and cusps, and in their details they may differ greatly from one substance to another. Nevertheless there are some features that all such frequency distributions share. One is the trivial one of normalization: since $g(\nu)d\nu$ is the fraction of modes with frequency between ν and $\nu + d\nu$ the total area under the $g(\nu)$ curve is 1. Another is that there is a maximum frequency, $\nu_{\max}$, with

$$g(\nu) \equiv 0 \quad \text{for } \nu \geq \nu_{\max}. \tag{4.1}$$

For LiF, as seen in Fig. 4.1, $\nu_{\max} = 19.6 \times 10^{12}$ Hz. The reason there is such a maximum frequency in the normal mode spectrum is that elastic waves of frequency ν have some speed of propagation c and wavelength λ, related by $\lambda \nu = c$, and there is a minimum wavelength λ set by the lattice spacing of the crystal: in the empty space between neighboring atoms there is nothing that can "wave." Because of the inverse relation between λ and ν a minimum λ translates into a maximum ν. In view of (4.1) the normalization of $g(\nu)$ may be expressed in either of the two alternative forms

$$\int_0^\infty g(\nu)d\nu = 1 \quad \text{or} \quad \int_0^{\nu_{\max}} g(\nu)d\nu = 1. \tag{4.2}$$

Another common feature of these $g(\nu)$ curves is that they show a rough division into two kinds of modes, those of low frequency and those of high frequency – although there is not a sharp dividing line between them. The low-frequency, long-wavelength modes constitute the "acoustic" branch of the spectrum. Their frequencies and wavelengths are those typical of sound waves; in those normal modes of vibration large groups of atoms move together, cooperatively, in the same direction. The high-frequency, short-wavelength modes constitute the "optical" branch. Their frequencies and wavelengths are those characteristic of infrared radiation, and they may be excited by and emit infrared radiation in optical spectroscopy. In those modes, typically, the motions of neighboring atoms are opposed; they move toward and away from each other instead of moving cooperatively in the same direction as they do in the acoustic modes.

Finally, there is one more feature that all such $g(\nu)$ functions have in common, one that is of central importance and will dominate much of the development in the later sections of this chapter: $g(\nu)$ is proportional to ν^2 when ν is small; i.e., $g(\nu)$ rises parabolically from 0 at the origin. The parabolic rise at $\nu = 0$ is clearly visible in Fig. 4.1. That universal feature of $g(\nu)$ at small ν is a consequence of the long wavelengths of those modes. Their wavelengths are so much greater than the linear dimensions of any characteristic structural features of the crystal that, to those modes, the crystal appears to be a structureless, elastic continuum; and all such elastic continua are much alike. In particular, as we shall see in the following section, the density $g(\nu)$ of their vibrational modes (number of modes per unit frequency range) is always proportional to ν^2; that is the "Rayleigh–Jeans law."

4.2 Rayleigh–Jeans law

Three kinds of waves may propagate in any given direction in an elastic body: longitudinal waves (which include sound waves), in which the displacements are in the direction of propagation, and the two kinds of transverse waves (shear waves), in which the displacements are in either of two mutually orthogonal directions both of which are perpendicular to the direction of propagation. These are transverse waves of two different "polarizations." For simplicity we shall assume the medium to be isotropic, so that the longitudinal waves have a

Fig. 4.2

speed of propagation c_ℓ and the two kinds of transverse waves have a common speed of propagation c_t, with c_ℓ and c_t both independent of the direction of propagation.

In either kind of vibration, at the point x, y, z in the body, at time t, there is a displacement $u(x, y, z, t)$ of the matter at that point from its equilibrium position. The displacement u is transverse to the direction of propagation of the wave, or in the direction of propagation, for the transverse and longitudinal waves, respectively. At any instant t, along a single line in the direction of propagation, the displacement u may have the form shown in Fig. 4.2. For the transverse waves this is both a *picture* of the wave – like a water wave or the waves on a vibrating string – and a *graph* of u as a function of position along the line of propagation; while for the longitudinal waves, for which the displacement u is along rather than perpendicular to the direction of propagation, the figure is only such a graph, not a picture of the wave.

For any one of the three kinds of wave, the longitudinal or either of the two polarizations of transverse wave, $u(x, y, z, t)$ satisfies the wave equation

$$\nabla^2 u = \frac{1}{c^2} \frac{\partial^2 u}{\partial t^2}, \qquad (4.3)$$

where c is either c_ℓ or c_t, respectively. This speed of propagation c is a physical property of the medium and differs from one medium to another. The normal modes, which we seek, are those for which $u(x, y, z, t)$ is periodic in t at any point x, y, z; i.e., they are those for which $u(x, y, z, t)$ is of the special form $U(x, y, z) \sin \omega t$, say, with $\omega = 2\pi \nu$ the radian frequency. Substituting this into (4.3), we find that U satisfies

$$\nabla^2 U = -\frac{\omega^2}{c^2} U. \qquad (4.4)$$

The independent normal modes of that type (longitudinal or either of the two transverse) correspond to a complete set of eigenfunctions U – solutions of (4.4) – those, say, obtained by imposing the condition $U = 0$ at the boundaries of the body. They are analogous to the complete set of solutions of the Schrödinger

4.2 Rayleigh–Jeans law

equation subject to appropriate boundary conditions in quantum mechanics. For simplicity, let us assume the body to be a cube of some edge-length L. Then, as may be verified by direct substitution, the solutions of (4.4) may be taken to be of the form

$$U = U_0 \sin(n_x \pi x/L) \sin(n_y \pi y/L) \sin(n_z \pi z/L)$$
$$(n_x, n_y, n_z = 1, 2, 3, \ldots) \quad (4.5)$$

with some constant amplitude U_0, provided

$$\frac{\pi^2}{L^2}(n_x^2 + n_y^2 + n_z^2) = \frac{\omega^2}{c^2} = \frac{4\pi^2 \nu^2}{c^2}; \quad (4.6)$$

i.e., provided

$$\nu = \frac{c}{2L}\sqrt{n_x^2 + n_y^2 + n_z^2}. \quad (4.7)$$

We have one normal mode of this type (longitudinal or either of the two transverse), of frequency ν given by (4.7), for each set of positive integers n_x, n_y, n_z. Different sets of n's correspond to different directions of propagation as well as to different frequencies or wavelengths; $n_x \pi/L$, etc., are the components of the "wave vector" **k**.

Exercise (4.1). Verify that Eq. (4.6),

$$\frac{\pi^2}{L^2}(n_x^2 + n_y^2 + n_z^2) = \frac{\omega^2}{c^2},$$

is the condition that (4.5),

$$U = U_0 \sin(n_x \pi x/L) \sin(n_y \pi y/L) \sin(n_z \pi z/L),$$

be a solution of (4.4),

$$\nabla^2 U = -\frac{\omega^2}{c^2} U.$$

Solution. ∇^2 means $\partial^2/\partial x^2 + \partial^2/\partial y^2 + \partial^2/\partial z^2$. We have

$$\frac{\partial^2}{\partial x^2}\sin(n_x \pi x/L) = (n_x \pi/L)\frac{\partial}{\partial x}\cos(n_x \pi x/L)$$
$$= -(n_x \pi/L)^2 \sin(n_x \pi x/L),$$

and the remaining factors in the postulated U are independent of x, so we conclude that that U satisfies

$$\frac{\partial^2 U}{\partial x^2} = -(n_x \pi/L)^2 U.$$

Similarly,

$$\frac{\partial^2 U}{\partial y^2} = -(n_y \pi/L)^2 U, \quad \frac{\partial^2 U}{\partial z^2} = -(n_z \pi/L)^2 U.$$

Therefore

$$\nabla^2 U = -\frac{\pi^2}{L^2}\left(n_x^2 + n_y^2 + n_z^2\right) U;$$

i.e., the postulated U satisfies

$$\nabla^2 U = -\frac{\omega^2}{c^2} U$$

provided

$$\frac{\pi^2}{L^2}\left(n_x^2 + n_y^2 + n_z^2\right) = \frac{\omega^2}{c^2},$$

as we wished to show.

In Fig. 4.3 we show the positive octant of a spherical shell of (inner or outer) radius $2Lv/c$ in a three-dimensional space the points of which have n_x, n_y, n_z as their coordinates. Each point with positive integer coordinates in this space corresponds to a normal mode of frequency v given by (4.7). The number of such normal modes with frequencies in the infinitesimal interval v to $v + \mathrm{d}v$ is then the number of such points with integer coordinates in the shell (the positive octant) shown in Fig. 4.3, when that shell, which is of radius $2Lv/c$, is taken to be of thickness $(2L/c)\mathrm{d}v$.

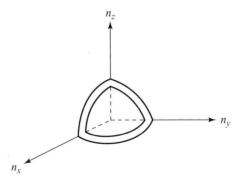

Fig. 4.3

4.2 Rayleigh–Jeans law

In thus speaking of a number of modes with frequencies in an infinitesimal interval we are supposing that the frequencies ν determined by (4.7) are practically infinite in number with nearly a continuum of possible values, and thus that there are many points with integer coordinates in the shell. But how can that be, when the shell appears to be of only infinitesimal thickness? The answer is that, as always in statistical mechanics, we are contemplating the thermodynamic limit; i.e., the limit in which the sample size, L in this instance, becomes macroscopically – essentially infinitely – large. If the frequency interval $d\nu$ is taken to be of order $1/\sqrt{L}$, say, it will then indeed, in a practical sense, be infinitesimal. The corresponding shell thickness, $(2L/c)d\nu$, will be of order $\sqrt{L}$, hence very large, yet much smaller than the radius, which is of order L. The volume contained between the inner and outer surfaces of the shell, which is the surface area times the thickness, hence $\frac{1}{8} \times 4\pi(2L\nu/c)^2(2L/c)d\nu$, will then also be very large, of order $L^{5/2}$, and will therefore contain a huge number of points with integer coordinates.

Since the volume contained between the inner and outer surfaces of the shell is large, and its thickness is also large, we may divide it into many cubes of unit volume, each cube having a point with integer coordinates at its center. There will be one such unit cube for each such point in the volume. Then, with only negligible corrections at the inner and outer surfaces of the shell where whole cubes cannot be made to fit, the volume in question will equal the number of such unit cubes that it contains; and so, conversely, the number of points in the volume that have integer coordinates will be equal to that volume. In the preceding paragraph we found the shell that corresponds to the normal modes of frequencies between ν and $\nu + d\nu$ to be of volume $\frac{1}{8} \times 4\pi(2L\nu/c)^2(2L/c)d\nu$. Therefore, this is also the number of those normal modes, which is thus $(4\pi V/c^3)\nu^2 d\nu$, where $V = L^3$ is the volume of the material sample. It was only for simplicity that we assumed the body to be a cube; this $(4\pi V/c^3)\nu^2$ would still have been the leading term in the expression for the number of modes per unit frequency interval, in the macroscopic limit as $V \to \infty$, for a sample of almost any shape that was not too bizarre.

Recall that this is the frequency distribution for only the one class of normal mode: longitudinal, in which case c means c_ℓ, or transverse with either of two possible polarizations, in which case c means c_t. Adding together the contributions from all three classes of normal mode, and letting $G(\nu)d\nu$ be the resulting total number of modes of all kinds that have their frequencies in the interval ν to $\nu + d\nu$, we thus find for an elastic continuum

$$G(\nu) = 4\pi V \left(1/c_\ell^3 + 2/c_t^3\right)\nu^2. \qquad (4.8)$$

This is the Rayleigh–Jeans law.

We remarked above that in the low-frequency limit $\nu \to 0$ the frequency-distribution function $g(\nu)$ of our model crystal would become that of an elastic continuum, for which we have now derived (4.8). Between $g(\nu)$ and $G(\nu)$ there is still a difference of normalization, however: $g(\nu) d\nu$ is the fraction of all the modes that have their frequencies in the given infinitesimal interval while $G(\nu)d\nu$ is the total number of such modes. Thus, if $f (= 3N$ with N again the number of atoms) is the total number of vibrational modes of the crystal sample, we have in the limit of low frequencies,

$$g(\nu) \sim 4\pi (V/f)(1/c_\ell^3 + 2/c_t^3)\nu^2 \quad (\nu \to 0). \tag{4.9}$$

This is the proportionality to ν^2 at low ν that we anticipated at the end of §4.1.

4.3 Debye theory of the heat capacity of solids

It is in accounting for the heat capacity of solids that this model – the ideal harmonic crystal – has had its most notable success. The theory is due to Debye.

We saw in Chapter 2, Eq. (2.29), that a harmonic vibration of frequency ν, when that vibration is independent of the other degrees of freedom of the system, contributes to the system's total partition function a factor

$$z_{\text{vib}} = \frac{1}{e^{\frac{1}{2}h\nu/kT} - e^{-\frac{1}{2}h\nu/kT}}. \tag{4.10}$$

That vibration contributes to the system's energy an amount [Eq. (2.33)]

$$\varepsilon_{\text{vib}} = -k\, d \ln z_{\text{vib}}/d(1/T) \tag{4.11}$$

and to the heat capacity an amount [Eq. (2.35)]

$$c_{\text{vib}} = d\varepsilon_{\text{vib}}/dT. \tag{4.12}$$

In §2.3 we carried out the indicated differentiations qualitatively and displayed the result in Fig. 2.3. The explicit result of the differentiation in (4.11) is in Exercise (2.3). The further differentiation in (4.12) then yields as the equation for the curve in Fig. 2.3,

$$c_{\text{vib}} = k \left(\frac{h\nu/kT}{e^{\frac{1}{2}h\nu/kT} - e^{-\frac{1}{2}h\nu/kT}} \right)^2. \tag{4.13}$$

The degrees of freedom in the present model are independent normal modes of vibration. Hence, each contributes additively to the total heat capacity the amount given by (4.13), according to its frequency ν. If again, as at the end of

4.3 Debye theory of the heat capacity of solids

§4.2, we let $f\,(=3N)$ be the total number of degrees of freedom, then $fg(\nu)d\nu$ is the total number of modes with frequencies in the interval ν to $\nu + d\nu$, so the total heat capacity of the model crystal is

$$C = fk \int_0^\infty \left(\frac{h\nu/kT}{e^{\frac{1}{2}h\nu/kT} - e^{-\frac{1}{2}h\nu/kT}} \right)^2 g(\nu) d\nu. \qquad (4.14)$$

[Again, as in (4.2), we may cut the integration off at $\nu_{\max}$.] Since the oscillators in this model are harmonic, thermal excitation affects the mean amplitudes of oscillation but not the mean positions of the atoms, so the volume of the crystal does not change with temperature. The model, then, is one in which the coefficient of thermal expansion is 0. In that case, as one learns in thermodynamics, $C_p - C_V = 0$; i.e., there is then no difference between C_V and C_p, so (4.14) may be taken to be either. Even for real solids C_V and C_p are usually not very different.

At high temperatures c_{vib} as given by (4.13) approaches the constant value k (cf. Fig. 2.3) and the factor that multiplies $g(\nu)$ in the integrand of (4.14) becomes 1. Then, because of the normalization (4.2), the heat capacity, according to (4.14), is just

$$C \sim fk = 3Nk \quad (T \to \infty), \qquad (4.15)$$

or $3R$ per mole. This is the law of Dulong and Petit, which was important historically in allowing the atomic weight of a solid element to be determined by measurement of its specific heat (heat capacity per unit mass).

The theoretical basis of the Dulong–Petit law had been long known just from the equipartition principle (§2.5). The real accomplishment of the Debye theory was its accounting for the heat capacity of solids in the opposite extreme of very low temperatures. For this it is convenient to let $h\nu/kT = x$ be a new variable of integration in (4.14), thus obtaining

$$C = (fk^2T/h) \int_0^\infty \left(\frac{x}{e^{x/2} - e^{-x/2}} \right)^2 g(kTx/h) dx. \qquad (4.16)$$

We then see that what the integral does at low T depends on the form of the frequency-distribution function g at small values of its argument. But that is just what we know from the Rayleigh–Jeans law and have expressed in (4.9). We thus find, at low temperatures,

$$C \sim (k^2T/h) \, 4\pi V \left(1/c_\ell^3 + 2/c_t^3\right)(kT/h)^2 \int_0^\infty \frac{x^4}{(e^{x/2} - e^{-x/2})^2} dx$$

$$(T \to 0). \qquad (4.17)$$

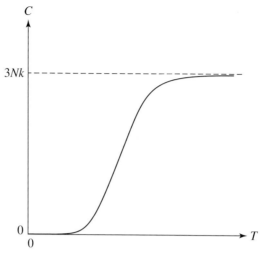

Fig. 4.4

The integral is a pure number and may be found in standard tables; it is $4\pi^4/15$. Thus,

$$C \sim (16\pi^5/15)k(kT/h)^3(1/c_\ell^3 + 2/c_t^3)V \quad (T \to 0). \quad (4.18)$$

It is the proportionality to T^3 at low T that is the essential result and that was needed for accord with experiment.

The full course of the curve of heat capacity vs. temperature is roughly determined by the two extremes we now know: C vanishing proportionally to T^3 at low T and approaching the constant $3Nk$ in the Dulong–Petit (high-T) limit. The full curve must then be qualitatively as shown in Fig. 4.4. Any $g(\nu)$ that has the property of being proportional to ν^2 at low ν will give C proportional to T^3 at low T and any $g(\nu)$ that is properly normalized will give C approaching $3Nk$ at high T. Thus, any $g(\nu)$ with just those two properties will give a C vs. T curve like that in Fig. 4.4.

Even a very primitive approximation to $g(\nu)$, as long as it has those two properties, can then be the basis of an interpolation formula that interpolates between the extremes of low and high T and gives a reasonable C vs. T curve for all T. Debye himself used the approximation

$$g(\nu) = \begin{cases} (3/\nu_D^3)\nu^2, & \nu < \nu_D \\ 0, & \nu > \nu_D, \end{cases} \quad (4.19)$$

4.3 Debye theory of the heat capacity of solids

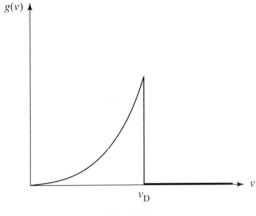

Fig. 4.5

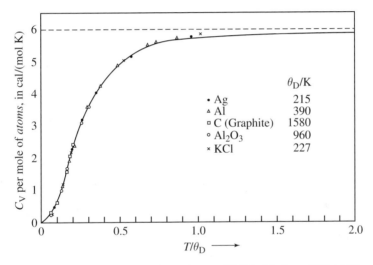

Fig. 4.6 After F. Seitz, *The Modern Theory of Solids* (McGraw-Hill, 1940).

shown graphically in Fig. 4.5. It has a single, adjustable, material-dependent parameter, the cutoff frequency ν_D. It is proportional to ν^2 at low ν in accord with the Rayleigh–Jeans law and is normalized so that the area under the $g(\nu)$ curve is 1. It is clearly only the roughest approximation to a realistic $g(\nu)$ such as that in Fig. 4.1, yet by yielding the correct behavior of the heat capacity at both low and high temperatures it produces heat-capacity curves (from the integral (4.14)) that are often close to those obtained by experiment at all T. This is shown in Fig. 4.6, where the heat capacity C_V per mole of *atoms* in the crystal, in cal/(mol K), is plotted against T/θ_D for several substances, with $\theta_D (= h\nu_D/k)$ chosen separately for each substance to give the best fit. The points are the data and the solid curve is the Debye approximation. The dashed line is the Dulong–Petit (classical equipartition) limit.

An historically earlier theory by Einstein took as a model a collection of harmonic oscillators all of the same frequency ν_E. That is equivalent to taking for $g(\nu)$ a delta function, $g(\nu) = \delta(\nu - \nu_E)$, centered at $\nu = \nu_E$. From (4.14) with $f = 3N$ it yields the approximation

$$C = 3Nk \left(\frac{h\nu_E/kT}{e^{\frac{1}{2}h\nu_E/kT} - e^{-\frac{1}{2}h\nu_E/kT}} \right)^2. \tag{4.20}$$

This is just all $3N$ oscillators making the same contribution, that in (4.13) with $\nu = \nu_E$, to the total. Since the delta function has unit area the normalization of $g(\nu)$ is right, so (4.20) goes to the right limit at high T. This $g(\nu)$ is not proportional to ν^2 at low ν, however, so (4.20) does not have the right T^3 behavior at low T. It does vanish at low T, but too rapidly: as $T^{-2} \exp(-h\nu_E/kT)$ instead of as T^3. Still, since it does vanish at low T and approaches the constant $3Nk$ at high T it, too, is qualitatively right over the whole temperature range (compare Figs. 2.3 and 4.6).

4.4 Black-body radiation

With only small modifications the foregoing theory may be adapted to give the density of radiant energy of any wavelength in a cavity ("black body") at thermal equilibrium at any temperature. This will be the Planck radiation law.

The radiation may be thought of as a collection of electromagnetic oscillators that have much in common with material oscillators. The electric and magnetic fields associated with light satisfy the same wave equations (4.3) and (4.4) as do the displacements u due to elastic waves in a material continuum. Historically, it was even thought for some time that radiation consisted of such excitations of a material, but ethereal, medium (the "ether"). The distribution in frequency of the standing electromagnetic waves in a cavity is given by the same Rayleigh–Jeans law (4.8) as for a material body, with two differences. First, electromagnetic waves are purely transverse, with no longitudinal component, so the term $1/c_\ell^3$ in (4.8) is absent here, and c_t is the speed of light, c:

$$G(\nu) = 8\pi V \nu^2 / c^3. \tag{4.21}$$

$G(\nu) d\nu$ is now the number of standing electromagnetic waves with frequencies in the interval ν to $\nu + d\nu$ in a cavity of volume V. Second, while (4.8) would

4.4 Black-body radiation

be just a low-ν approximation for a real material – that being the limit in which the material may be thought of as an elastic continuum – (4.21) for radiation is exact at all ν. It is as though the "ether" were a true continuum with no microscopic structure revealed at high frequencies (short wavelengths).

The average energy $\varepsilon(\nu)$ of any one such electromagnetic oscillator of frequency ν, in thermal equilibrium at temperature T, may as usual be found from

$$\varepsilon(\nu) = -k\, \mathrm{d}\ln z/\mathrm{d}(1/T), \tag{4.22}$$

as in (4.11), where z is the partition function of the oscillator. Here, however, z is a little different from that for a material oscillator given in (4.10). That is because (4.10), which is (2.29) of Chapter 2, was derived with the vibrational energy levels $(v + \tfrac{1}{2})h\nu$ $(v = 0, 1, 2, \ldots)$, which include the zero-point energy $\tfrac{1}{2}h\nu$. Electromagnetic oscillators, however, unlike material ones, have no zero-point energy. Their energy levels, as originally surmised by Planck and Einstein, are just $nh\nu$ $(n = 0, 1, 2, \ldots)$, corresponding to n "photons" of frequency ν. If we re-derive (2.29) with these energy levels the factor $\exp(-\tfrac{1}{2}h\nu/kT)$ is no longer present in the next-to-last line of (2.29), and z becomes just

$$z = \frac{1}{1 - \mathrm{e}^{-h\nu/kT}}. \tag{4.23}$$

Then from (4.22),

$$\begin{aligned}\varepsilon(\nu) &= \frac{h\nu\, \mathrm{e}^{-h\nu/kT}}{1 - \mathrm{e}^{-h\nu/kT}} \\ &= \frac{h\nu}{\mathrm{e}^{h\nu/kT} - 1}.\end{aligned} \tag{4.24}$$

Equation (4.24) gives the average energy of an oscillator of frequency ν, while $G(\nu)\mathrm{d}\nu$ is the number of such oscillators with frequencies in the range ν to $\nu + \mathrm{d}\nu$. Therefore, if $E(\nu)\mathrm{d}\nu$ is the total radiant energy in that infinitesimal frequency range, at equilibrium at temperature T in the cavity of volume V, we have from (4.21) and (4.24),

$$\begin{aligned}E(\nu) &= G(\nu)\varepsilon(\nu) \\ &= \frac{8\pi V h (\nu/c)^3}{\mathrm{e}^{h\nu/kT} - 1}.\end{aligned} \tag{4.25}$$

This is Planck's radiation law.

Planck's law (4.25) is in accord with experimental measurements of $E(\nu)$ as inferred from spectral analysis of the radiation emitted through a pinhole in the

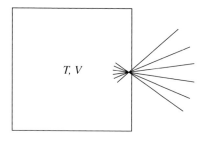

Fig. 4.7

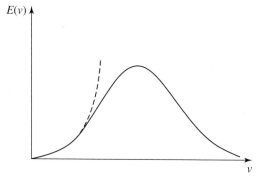

Fig. 4.8

wall of a cavity that is held at a fixed temperature T (Fig. 4.7). The function $E(\nu)$ is displayed in Fig. 4.8. The dashed curve in the figure is the high-temperature approximation obtained by taking for $\varepsilon(\nu)$ its high-temperature, equipartition value kT, as follows from (4.24) with $T \to \infty$ [cf. (2.34) of Chapter 2].

That approximation, $E(\nu) = kT G(\nu) = 8\pi V kT \nu^2/c^3$, is Rayleigh's radiation law. It is correct in the limit of low frequencies, which is the same as the limit of high temperatures since T occurs only in the combination ν/T in the Planck law (4.25). It fails at high frequencies because it has $E(\nu)$ continuing to increase proportionally to ν^2 instead of reaching a maximum and then decreasing. It would lead to an infinite instead of finite total energy (area under the $E(\nu)$ curve). That divergence due to the incorrect behavior at high frequencies was known in the early literature as the "ultraviolet catastrophe." The empirical observation by Planck that the experimentally measured $E(\nu)$ was fit by (4.25) was the beginning of quantum mechanics.

5
The third law

5.1 Nernst heat theorem in thermodynamics

The "third law of thermodynamics," unlike the first and second laws, does not introduce any new functions of state but rather asserts a universal behavior of the entropy at low temperatures. The most useful form of the third law is the Nernst "heat theorem," a simplified statement of which is that the entropies of substances approach 0 at the absolute zero of temperature. This statement, which needs much qualification, is explained more fully in this section (a reminder of what one may have learned in thermodynamics), and then its statistical mechanical basis is presented in §5.2.

Suppose a system undergoes a change in its thermodynamic state at a fixed temperature T (isothermal change), and that the accompanying entropy change is ΔS_T. The subscript reminds us that the entropy change in general depends on the temperature T at which the process occurs. The process could be a chemical reaction, a change of physical state (melting, for example, or transformation between two different crystalline forms), an isothermal expansion or compression, etc. We ask only that the system in both its initial and final states be in a condition of complete internal equilibrium so that it would be possible in principle to carry out the process reversibly (however it may actually be carried out) and so that the system has well defined values of its thermodynamic functions both initially and finally.

The system could well be metastable rather than absolutely stable in its initial or final state provided the other conditions are satisfied. An example is the freezing of supercooled water at 1 atm and -5 °C. This is not an equilibrium process: the Gibbs free energy of ice at 1 atm and -5 °C is lower than that of water at that pressure and temperature. The process may nevertheless be made to occur reversibly; for example, by reversibly heating the water to 0 °C, letting it undergo its equilibrium transformation to ice, and then reversibly cooling the ice to $-5\,°C$, all at 1 atm. A chemical reaction in which the reactants and products are not in chemical equilibrium with each other would be another example, provided the reactants and products separately are in well defined thermodynamic states.

Since we shall ultimately be considering what happens in the limit of very low temperatures we are almost always talking about processes in which the system is a solid in both its initial and final states (although a process involving helium, which below 25 atm is stable as a liquid down to $T=0$, could be an exception). Thus, an example of a chemical reaction that we might consider is that of elemental lead and sulfur to form lead sulfide,

$$Pb(s) + S(s) \to PbS(s), \qquad (5.1)$$

and an example of a change of crystal structure might be the conversion of monoclinic to rhombic sulfur,

$$S(mon) \to S(rhomb). \qquad (5.2)$$

Neither of these, in general, would be an equilibrium change, for at the temperature T and the pressure at which it occurs the enthalpy change ΔH_T would not in general equal $T\Delta S_T$, but they are nevertheless examples of the kinds of processes we consider. Another example might be the isothermal compression of a solid, say ammonium iodide, from a pressure p_1 to a pressure p_2,

$$NH_4I(p_1) \to NH_4I(p_2). \qquad (5.3)$$

Exercise (5.1). Invent a reversible path that starts and ends at some temperature T_1, for the transformation of $Pb(s) + S(s)$ into $PbS(s)$.

Solution. Holding fixed whatever pressures the reactants and products are at, there is (in principle) a temperature T_2 at which the reactants and products are in equilibrium; viz., the solution T_2 of the equation $\Delta H_T = T\Delta S_T$, where ΔH_T and ΔS_T are the enthalpy and entropy changes in the reaction at temperature T. Then reversibly heat (or cool) the reactants from T_1 to T_2, allow the reaction to occur reversibly at T_2, and then reversibly cool (or heat) the products from T_2 to T_1. Alternatively, at fixed $T=T_1$ there is (in principle) a pressure at which $Pb(s) + S(s)$ and $PbS(s)$ are in equilibrium; it is the solution p of $\Delta H_{T_1}(p) = T_1 \Delta S_{T_1}(p)$ (ΔH_{T_1} and ΔS_{T_1} being in general both pressure dependent). Then reversibly and isothermally compress (or expand) the reactants to that chemical-equilibrium pressure p at the temperature T_1, allow the reaction to occur reversibly at fixed p and T_1, and then reversibly and isothermally expand (or compress) the product to its original pressure.

The third law of thermodynamics asserts as an empirical fact that as the temperature T at which such an isothermal process occurs approaches 0, so does the accompanying entropy change ΔS_T:

$$\lim_{T \to 0} \Delta S_T = 0. \tag{5.4}$$

This is the Nernst heat theorem.

It says that the entropies in the initial and final states become equal at $T = 0$. This is true whether the change be merely one of physical state (crystal structure or density) or a change of chemical identity (chemical reaction). For all such states that are in principle transformable one into the other by a reversible process, physical or chemical, one may arbitrarily assign a value of the entropy in one of them at $T = 0$ and then, by (5.4), that will also be the value of the entropy in all the others at $T = 0$. By a generally accepted convention, that universal value of the entropy at $T = 0$ is taken to be 0. For substances that are not transformable into each other by chemical reaction that convention may be, and is, applied separately to each. Thus, we arrive at the statement that the entropies of substances in complete internal equilibrium, i.e., in well defined thermodynamic states, are all 0 at $T = 0$; but we must understand that that 0 for the value of the entropy is an arbitrary convention, not a law of nature. The law of nature is (5.4), which is a law about entropy changes. Only such differences are measurable physical quantities.

We saw in Chapter 1 that the arbitrariness in the zero of entropy in thermodynamics is reflected in statistical mechanics; we arbitrarily set $\phi = 0$ in (1.15) and thus took the Helmholtz free energy to be related to the partition function by (1.16). We shall see in the next section that that convention in statistical mechanics is precisely the one that makes the entropies of substances 0 at $T = 0$, and so is consistent with the usual convention in thermodynamics. More importantly, §5.2 gives the statistical mechanical, hence microscopic, basis of the third law. It shows that law to be a consequence of quantum mechanics' manifesting itself strongly at low temperatures.

5.2 Third law in statistical mechanics

We saw in Chapter 1, Eq. (1.26), that in a thermodynamic state of energy U a system's entropy S is related to its density of states $W(E)$ evaluated at $E = U$, by

$$S(U) = k \ln[W(U)\Delta E], \tag{5.5}$$

where ΔE is some energy interval the logarithm of which is sub-extensive (so that it hardly matters to S just what ΔE is). The function W depends also on the volume and chemical composition of the system, and so does S, but here we emphasize the energy dependence. It should also be recalled that (5.5) was derived from the relation (1.16) between the Helmholtz free energy A and the partition function Z, which required choosing the arbitrary ϕ in (1.15) to be 0. Otherwise, S as obtained from $S = -\partial A/\partial T$ [Eq. (1.17)] would have contained the additional term $-\phi$, the value of which is arbitrary, and (5.5) would have read $S = k \ln[W(U)\Delta E] - \phi$.

We remarked earlier (§1.3) that the extensivity of S means that $W(E)\Delta E$ at any energy E is of the order of the exponential of a quantity that is itself of the order of the number of molecules or number of mechanical degrees of freedom, N, in the system, and that almost all of that comes from $W(E)$ since ΔE contributes negligibly. From the inverse of (5.5) we have for the density of states $W(E)$ at any energy E,

$$W(E) = (\Delta E)^{-1} e^{S(E)/k}; \tag{5.6}$$

i.e., the exponent in question is the extensive $S(E)/k$.

The crux of the third law is now the remark that the density of the quantum states of any real, macroscopic system is especially low in the neighborhood of the ground state – so low, that while $W(E)$ is the exponential of something macroscopically large, $O(N)$, at all energies E above the ground-state energy E_0, at E_0 itself W is the exponential of something sub-macroscopically large. If we use the symbol $o(N)$ (as distinct from $O(N)$) to mean a quantity – like $\sqrt{N}$ or $\ln N$, for example – that is smaller in magnitude than N, so that $o(N)/N \to 0$ in the thermodynamic limit, then $\ln[W(E_0)\Delta E] = o(N)$. This is depicted schematically in Fig. 5.1, where the levels are shown becoming sparse as E decreases toward E_0; i.e., rapidly *increasing* in density as E *increases* from E_0. The discreteness of the levels in Fig. 5.1 is schematic only. One must always imagine being in the thermodynamic limit in which there is a continuum of levels with a density $W(E)$. It will be seen later, though, that the particularly low density $W(E_0)$ in the macroscopic limit is indeed a reflection of the relative sparseness of the low-lying levels of a finite system, for which the levels are discrete.

From (5.5) or (5.6), this uniquely low density of states in the neighborhood of the ground state means that, while $S(U)$ at all energies U macroscopically above the ground-state energy E_0 is extensive, hence macroscopically large,

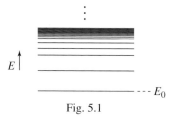

Fig. 5.1

5.2 Third law in statistical mechanics

the entropy approaches something sub-extensive as $U \to E_0$. In that sense the entropy (after our having made the conventional choice for its arbitrary additive constant) approaches "0" as the energy U approaches the ground-state energy E_0 of the system. More precisely, it means that the entropy per molecule or per degree of freedom, S/N, or the entropy density S/V, *after* we have gone to the thermodynamic limit of infinite N and V, approaches 0 as the energy per molecule or per degree of freedom, U/N, or the energy density U/V, *then* approaches that of the ground state. Note the order in which the limits are taken: first the thermodynamic limit of a macroscopically (in principle, infinitely) large system, and only then the approach to the ground state. Had the order been reversed, so that the ground state of a system of finite size was being approached, we would have found discrete energy levels with a ground state that was separated in energy from that of the first excited state by a measurable gap, and that would have given a false impression of the microscopic origins of the third law. The third law is not about the discreteness of the levels but about their unique paucity near the ground state.

The origins of this thinness of the level density near the ground state can be seen clearly in a simple model such as the Einstein model of a harmonic crystal (§4.3): a collection of a macroscopically large number, say N, of harmonic oscillators all with a common frequency ν. If we take energy 0 to correspond to the minimum of potential energy, then the ground state of the system, which is that in which each of the N oscillators has only its zero-point energy $\frac{1}{2}h\nu$, is at $E_0 = \frac{1}{2}Nh\nu$ (Fig. 5.2). That ground state is non-degenerate: degeneracy $g_0 = 1$. The first excited state of the system is at $E_1 = \frac{1}{2}(N+2)h\nu$; it is that state in which one of the oscillators has one quantum of excitation, of energy

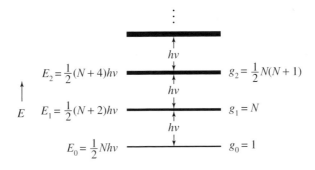

Fig. 5.2

$h\nu$, while the $N-1$ others are still in their ground states. Any of the N can be the one that is excited, so this first level above the ground level is N-fold degenerate: $g_1 = N$. The next level is at $E_2 = \frac{1}{2}(N+4)h\nu$, corresponding to two quanta of excitation in the system. The two quanta can be in different ones of the N oscillators, which can happen in $\frac{1}{2}N(N-1)$ ways, or can both be in the same oscillator, which can happen in N ways, so the total degeneracy is $g_2 = \frac{1}{2}N(N-1) + N = \frac{1}{2}N(N+1)$; and so on. We see an extremely rapid increase of degeneracy with increasing energy, or, going in the other direction, an extremely rapid decrease with decreasing energy. If now we imagine a more realistic model in which these oscillators are coupled by anharmonic couplings, those degeneracies would be split, the split levels would overlap to fill in the gaps in the spectrum, and the result, when N was large, would be a near continuum of levels distributed with a density that decreased rapidly with decreasing energy. Indeed, the only reason that the degeneracies shown in Fig. 5.2, which as energy increases are of order N^0, N^1, N^2, ..., are as *small* as they are, is because we are there looking only at microscopic excitation energies $h\nu$, $2h\nu$, ... At any macroscopic energy, $O(N)$, above the ground state, the density of levels in the system with anharmonic couplings would be of the order of $\exp[O(N)]$; and would then decrease rapidly on approach to the ground state, as is already apparent in the simplified scheme in Fig. 5.2.

We have thus seen that the entropy $S(U)$ vanishes as U approaches the ground-state energy E_0, and why and in what sense it does so. Since the thermodynamic energy U approaches E_0 as the temperature T approaches 0, this means that the entropy considered as a function of T vanishes in the limit $T \to 0$. That is the statement of the third law to which we came at the end of §5.1.

A system might well have a ground state with an unsplit degeneracy of the order of $\exp[O(N)]$ – for example, the system might contain N spin-$\frac{1}{2}$ nuclei interacting so weakly with each other and with the other degrees of freedom of the system that the ground state is in effect 2^N-fold degenerate because of the nuclear spin degeneracy; or, even if there is a slight splitting of those nuclear levels due to weak interactions, the resulting energy-level spread might still be much less than kT down to the lowest practically attainable T, and so be unnoticeable. Then, if the entropy is thought to approach 0 as $T \to 0$ when those nuclear degrees of freedom are ignored, it will instead have to be taken to approach $k \ln 2^N = Nk \ln 2$ as $T \to 0$ when that extra degeneracy is taken into account – for the extra degeneracy will have introduced an extra factor 2^N into the level density $W(E)$. The density of states near $E = E_0$ would not be as small as we had earlier imagined. Indeed, there may be some as yet undiscovered degrees of freedom associated with nuclei or electrons that would lead us to assign a still greater degeneracy to the ground state and so an even greater value to the limiting entropy at $T = 0$. What, then, is the status of the statement that

$S \to 0$ as $T \to 0$? The answer is that the very weakness of the interactions with each other and with the other degrees of freedom of the system that makes those extra degrees of freedom indiscernible makes them remain unaffected by any change in the system's thermodynamic state. The extra degeneracy is the same in any state and so contributes nothing to any measurable ΔS. We may therefore, if we wish, ignore those degrees of freedom in evaluating the partition function Z, continue to adopt the convention $\phi = 0$ in (1.15), and so continue to conclude that on such a scale the limiting entropy of a system in complete internal equilibrium at $T = 0$ is 0.

5.3 Comparison with experiment

One may measure experimentally the entropy change ΔS in the process pictured in Fig. 5.3. In some stages of the process a phase – solid, liquid, or gas – is simply being heated from a temperature T_1 to a temperature T_2 (at a fixed pressure, say). The entropy increment for each such stage is just

$$\int_{T_1}^{T_2} \frac{C_p}{T} \, dT, \tag{5.7}$$

by virtue of the thermodynamic identity $C_p = T(\partial S/\partial T)_p$, so the entropy increment is obtained from measurements of the constant-pressure heat capacity of the phase. At other stages of the process pictured in Fig. 5.3 phase transitions – changes of crystalline form, melting, or evaporation – occur. The entropy increment in each such stage is $\Delta H/T$ where ΔH is the enthalpy change (latent heat) in the transition and T the temperature at which, for the given pressure, the phase transformation is an equilibrium one. Those entropy increments, then, are obtained from measurements of latent heats and transition temperatures.

There is a practical limit to how low one can go in temperature, but if the C_p measurements for use in (5.7) are carried to low enough temperature the remaining contribution down to $T = 0$ can be reliably obtained by extrapolation

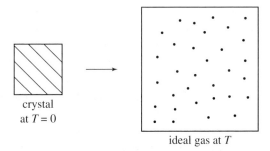

Fig. 5.3

76   5 *The third law*

using Debye's T^3 law (§4.3) as a guide. That is of little consequence because the contribution to the integral in (5.7) from very low temperatures is small, the integrand being proportional to T^2 at low T. There is likewise a practical limit to how nearly ideal the gas in the final state in Fig. 5.3 can be. Typically, the gas is taken to be at 1 atm, and then a small theoretical correction, based on ideas and methods related to those we shall study in the next chapter, is applied to remove the effects of non-ideality at that pressure; but that correction, too, is very small.

The entropy increment in the process in Fig. 5.3 can also be calculated theoretically. The entropy in the final, ideal-gas state may be calculated from the partition functions we evaluated in Chapter 2. Those, we have seen, are on a scale on which the entropy in the initial state – the crystal at $T = 0$ – is 0, if that is a state of complete internal equilibrium. Thus, if our theory, including the statistical mechanical analysis of the origins of the third law in §5.2, is correct, the S_{theor} we calculate for the ideal gas from the partition functions in Chapter 2, hence essentially from spectroscopic data, should be identical with the ΔS_{expt} obtained by experiment for the process in Fig. 5.3, from measurements of heat capacities and latent heats. The comparison of S_{theor} with ΔS_{expt} would be a rigorous test of statistical mechanical theory.

Exercise (5.2). Calculate the molar entropy S_{theor} of CH_4 at 25 °C and 1 atm, as an ideal gas, and compare with the experimentally determined ΔS_{expt}, which, corrected for slight deviations from ideality, is 44.5 cal/(mol K). [Methane is a spherical-top molecule; i.e., its three moments of inertia are all equal, with a common $\theta_{rot} (= h^2/8\pi^2 I k) = 7.54$ K and a symmetry number $\sigma = 12$. Its vibrational mode of lowest frequency (actually a degenerate triple of such modes) has $\theta_{vib} = 1870$ K.]

Solution. The molar entropy of translation S_{trans} is given by the Sackur–Tetrode equation, (2.26),

$$S_{trans} = R \ln \left[\left(\frac{2\pi m k T}{h^2} \right)^{3/2} \frac{kT}{p} e^{5/2} \right].$$

Since $\theta_{rot} = 7.54$ K $\ll 298$ K we may use the classical z_{rot} in (2.51) with $I_1 = I_2 = I_3 = I$. Thus, $z_{rot} = \frac{1}{12}\sqrt{\pi}(T/\theta_{rot})^{3/2}$. The rotational contribution to the Helmholtz free energy per molecule is $a_{rot} = -kT \ln z_{rot}$, and to the energy per molecule is $\varepsilon_{rot} = (3/2)kT$, and so to the entropy per molecule is $s_{rot} = (\varepsilon_{rot} - a_{rot})/T = (3/2)k + k \ln z_{rot}$; or, per mole,

$$S_{rot} = R \ln \left[\frac{1}{12} \sqrt{\pi} (eT/\theta_{rot})^{3/2} \right].$$

5.3 Comparison with experiment

Since all the θ_{vib} are at least as high as 1870 K, and so are much higher than 298 K, each z_{vib} may be taken to be $\exp(-\theta_{vib}/2T) = \exp(-\frac{1}{2}h\nu/kT)$, as in (2.31) with (2.30); each $a_{vib} = -kT \ln z_{vib} = \frac{1}{2}h\nu$; and each $\varepsilon_{vib} = \frac{1}{2}h\nu$. Then $s_{vib} = (\varepsilon_{vib} - a_{vib})/T = 0$: vibrations with $\theta_{vib} \gg T$ contribute negligibly to the entropy. With numerical values $R = 1.9872$ cal/(mol K), $m = (16.04/6.0225 \times 10^{23})$ g, $p = 1$ atm $= 1.01325 \times 10^6$ dyne/cm², $k = 1.3806 \times 10^{-16}$ erg/K, $h = 6.6262 \times 10^{-27}$ erg s, $T = 298.15$ K, and $\theta_{rot} = 7.54$ K, we calculate $S_{trans} = 34.26$ cal/(mol K) and $S_{rot} = 10.14$ cal/(mol K). Then $S_{theor} = 44.40$ cal/(mol K), in virtually exact agreement with the $\Delta S_{expt} = 44.5$ cal/(mol K). Indeed, the small vibrational contribution, calculated exactly, is 0.10 cal/(mol K), which makes the agreement even closer.

Exercise (5.3). The equilibrium internuclear distance in N_2 is 1.094 Å and the vibration frequency of $^{14}N_2$ is 7.07×10^{13} s^{-1}. The ground electronic state of N_2 is non-degenerate. Calculate the molar entropy of N_2 as a hypothetical ideal gas at 1 atm and 25 °C (with the usual convention for the zero of entropy), and compare with the experimental value (corrected for non-ideality), 191.5 J/(mol K).

Solution. With m the mass of the N atom the mass of the molecule is $2m$ and the moment of inertia $I = \mu r^2 = \frac{1}{2}mr^2$ with r the equilibrium internuclear distance. (Here we use r rather than R so as not to confuse it with the gas constant.) From the Sackur–Tetrode equation [cf. Exercise (5.2)],

$$S_{trans} = R \ln \left[\left(\frac{2\pi \cdot 2mkT}{h^2} \right)^{3/2} \frac{kT}{p} e^{5/2} \right].$$

$\theta_{vib} = h\nu/k = 3390$ K $\gg 298$ K so $S_{vib} = 0$, essentially. $\theta_{rot} = h^2/(8\pi^2 Ik) = h^2/(4\pi^2 mr^2 k) = 3$ K $\ll 298$ K, so

$$z_{rot} = \frac{8\pi^2 IkT}{2h^2} \quad [\sigma = 2;\ \text{Eq. (2.40)}]$$

$$a_{rot} = -kT \ln z_{rot}$$

$$\varepsilon_{rot} = kT \quad [\text{Eq. (2.48)}]$$

$$s_{rot} = \frac{\varepsilon_{rot} - a_{rot}}{T} = k \ln(ez_{rot})$$

$$S_{rot} = R \ln \left(e \frac{8\pi^2 IkT}{2h^2} \right) = R \ln \left(e \frac{2\pi^2 mr^2 kT}{h^2} \right).$$

Then

$$S = S_{rot} + S_{trans} \quad (+ S_{vib}, \text{ which is } \simeq 0)$$

$$= R \ln \left[\left(\frac{4\pi mkT}{h^2} \right)^{3/2} \frac{kT}{p} e^{7/2} \frac{2\pi^2 mr^2 kT}{h^2} \right].$$

With $m = (14/6.022 \times 10^{23})$ g, $\quad r = 1.094 \times 10^{-8}$ cm,
$T = 298.15$ K, $\quad p = 101{,}325$ Pa $= 1.01325 \times 10^6$ g/(cm s^2),
$h = 6.626 \times 10^{-34}$ J s $= 6.626 \times 10^{-27}$ erg s,
$k = 1.3807 \times 10^{-23}$ J/K $= 1.3807 \times 10^{-16}$ erg/K, and
$R = 8.3145$ J/(mol K), we calculate

$$S = 191.4 \text{ J/(mol K)},$$

essentially in agreement with experiment.

More generally than the process pictured in Fig. 5.3, we might consider that in Fig. 5.4, where the initial state is that of a pure substance in crystalline form at $T = 0$ while the final state may be any equilibrium state of that substance – solid, liquid, or gas. Again ΔS_{expt} may be measured, as before, and with the entropy in the initial state at $T = 0$ taken to be 0 one would assign the measured ΔS_{expt} as the entropy in the final state. In principle that is also the entropy we would calculate theoretically from the partition function, although that might be difficult for any state of the substance other than the dilute gas. That entropy – whether measured or calculated – is often called the "third-law" entropy or, misleadingly, the "absolute" entropy of the substance in that state. It is what is frequently tabulated as the substance's "standard" entropy when the state (the final state in Fig. 5.4) is some arbitrarily chosen standard state – e.g., 1 atm

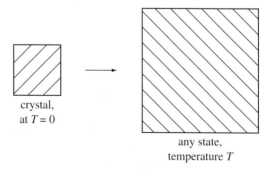

Fig. 5.4

and 25 °C with the substance in some specified form, usually that in which it is most stable at that pressure and temperature.

The agreement between theory and experiment illustrated in Exercises (5.2) and (5.3) is typical of that for most simple, carefully studied substances, but there are some well known and frequently quoted exceptions, among them N_2O, CO, and H_2O. For N_2O the measured ΔS_{expt} at 1 atm and 298 K, corrected for non-ideality, is 51.44 cal/(mol K) while that calculated theoretically is 52.58 cal/(mol K), which is greater by 1.14 cal/(mol K). For CO the discrepancy between theory and experiment is nearly the same, 1.11 cal/(mol K), and for H_2O it is 0.82 cal/(mol K) – always in the same direction, $\Delta S_{\text{expt}} < S_{\text{theor}}$. For such cases we may say that the entropy has been found not to vanish at $T = 0$ (on the entropy scale we have adopted) but to have some positive residual value there; the residual entropy of N_2O is 1.14 cal/(mol K), etc.

That means that in each of those cases there is some unresolved degeneracy of the ground state of the crystal that makes the density of states $W(E_0)$ greater, by a factor equal to the exponential of a quantity $O(N)$, than we had expected. This extra degeneracy in crystalline N_2O and CO is usually ascribed to the randomness with which each linear N_2O or CO molecule chooses either of two possible orientations in the crystal. The ordered state in which each molecule has its proper orientation is only slightly lower in energy than the disordered state in which the two orientations occur randomly, so at temperatures well above $T = 0$ the equilibrium state of the crystal is the disordered one. At lower temperatures the equilibrium state would be partially ordered, and at $T = 0$ it would be completely ordered; but at those low temperatures the kinetics of the ordering would be very slow because the activation energy for molecular reorientation would then be much greater than kT. Thus, except perhaps for some slight annealing, the disorder characteristic of the higher temperatures would be "frozen in" at the lower. If each of N molecules may have either of 2 orientations then there are 2^N possible states, so an extra factor of 2^N in the density of states near the ground state, so a residual entropy of $k \ln 2^N = Nk \ln 2$, or, per mole, $R \ln 2 = 1.38$ cal/(mol K). If this is indeed the origin of the residual entropy in N_2O and CO it means that there must have been some slight annealing (ordering) at the lower temperatures, since the measured residual entropies are a little less than this theoretical value.

The measured residual entropy of ice, 0.82 cal/(mol K), has been explained by Pauling [L. Pauling, *The Nature of the Chemical Bond*, Cornell Univ. Press (1945) pp. 301–303]. Each O atom in ice is surrounded tetrahedrally by four others, and between each O and each one of its four neighboring O's is

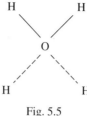

Fig. 5.5

an H. Two of these H's are closer to the central O than to the neighboring O's (corresponding to a discrete H_2O molecule), and two are closer to the neighboring O's. That is the hydrogen-bonded structure of ice: each H atom is on a bond (hydrogen bond) between a pair of neighboring O's. Figure 5.5 shows, schematically, one central O and its four neighboring H's.

If there were not the restriction of two close and two distant H's at each O, each H would have the choice of two sites, and an ice crystal consisting of N H_2O molecules would have a 2^{2N}-fold ($=4^N$-fold) degenerate ground state. Each O in that case would have been found in one of sixteen possible states (but not independently of its neighboring O's, so corresponding to only 4^N, not 16^N, states of the crystal); one of the sixteen being representable as H_4O^{++} (all four H's close to the central O), four as H_3O^+, six as H_2O, four as HO^-, and one as $O^=$. But of those sixteen, only the six that are representable as H_2O, for which there are two close H's and two distant H's at each O, actually occur. Thus, at each O only the fraction $6/16 = 3/8$ of the states that would have been possible if each of the four H's had been free to choose its position independently of the other three, actually occur. Thus, for the system as a whole only the fraction $(3/8)^N$ of the 4^N states in which each H is free to choose either of two positions, actually occur: the degeneracy is $(3/8)^N \times 4^N = (3/2)^N$. Then the residual entropy is $k \ln (3/2)^N = Nk \ln(3/2)$, or, per mole, $R \ln(3/2) = 0.81$ cal/(mol K), in essential agreement with experiment.

6

The non-ideal gas

6.1 Virial coefficients

The familiar ideal-gas law

$$pV = nRT (= NkT) \qquad (6.1)$$

would hold strictly, as we know (Chapter 2), only if the molecules did not interact with each other. In practice that means it holds accurately only for dilute gases – for then the distance between neighboring molecules is on average very large, so that the forces they exert on each other are generally weak, while those instances when they do interact strongly (when they "collide") are rare. We therefore expect that the ideal-gas law (6.1) is only the low-density limit, $N/V \to 0$, of a more general gas law in which p/kT is N/V to leading order in the small quantity N/V but in which there are successively smaller correction terms of order $(N/V)^2$, $(N/V)^3$, etc.

Such a representation of the equation of state is the *virial series*,

$$\begin{aligned} p/kT &= N/V + B(N/V)^2 + C(N/V)^3 + \cdots \\ &= (N/V)[1 + BN/V + C(N/V)^2 + \cdots]. \end{aligned} \qquad (6.2)$$

(In classical thermodynamics the number density is usually expressed as the number of moles per unit volume, n/V, rather than as the number of molecules per unit volume, N/V. The corresponding coefficients $B, C, \ldots$ then differ from the present ones by factors that are powers of Avogadro's number.) The name virial series comes from the so-called virial theorem of mechanics, which, historically, provided the first route to the expansion (6.2). Our own derivation, in §6.3, will be from a different direction.

The pressure p is generally some function $p(T, N/V)$ of temperature T and number density N/V. Equation (6.2) is an expansion of $p(T, N/V)/kT$ in powers of N/V in which the coefficients $B, C, \ldots$ are then functions of

T. These temperature-dependent coefficients $B(T)$, $C(T)$, ... are termed *virial* coefficients: B is the second virial coefficient, C the third, etc. (The "first" virial coefficient is just 1, but nobody refers to it as a virial coefficient.)

The first discernible deviations from ideality with increasing density are those arising from the first correction term, BN/V, in (6.2). In a gas that is dilute but not so dilute that the ideal-gas law $p/kT = N/V$ would be accurate enough, the single correction term BN/V is generally sufficient. Thus, of the virial coefficients B, C, etc., it is usually only the second virial coefficient, $B(T)$, that is of any importance, and it is always the most important. It is from the temperature dependence of $B(T)$ that one determines the Boyle temperature and the low-pressure Joule–Thomson inversion temperature of the gas.

Exercise (6.1). Calculate the second virial coefficient $B(T)$ implied by the van der Waals equation of state $(p + an^2/V^2)(V/n - b) = RT$, and then obtain the Boyle temperature T_B from $B(T_B) = 0$ and the low-pressure Joule–Thomson inversion temperature as the solution T_i of $d[B(T)/T]/dT = 0$. (These concepts of the Boyle temperature and Joule–Thomson inversion temperature and their quoted relation to the second virial coefficient are known from classical thermodynamics.)

Solution.

$$p = RT/(V/n - b) - an^2/V^2$$
$$= (nRT/V)/(1 - bn/V) - an^2/V^2$$
$$= (nRT/V)[1 + bn/V + b^2n^2/V^2 + \cdots] - an^2/V^2,$$

where we have made use of the expansion $1/(1-x) = 1 + x + x^2 + x^3 + \cdots$. Then

$$p = (nRT/V)[1 + (b - a/RT)n/V + b^2n^2/V^2 + \cdots].$$

With $nR = Nk$ and $n = N/N_0$, where N_0 is Avogadro's number, we find, on comparing this with (6.2), that

$$B(T) = N_0^{-1}(b - a/RT).$$

We thus find the Boyle temperature $T_B = a/Rb$; while the low-pressure Joule–Thomson inversion temperature T_i is the solution of $0 = d(b/T - a/RT^2)/dT = -b/T^2 + 2a/RT^3 = (-b + 2a/RT)/T^2$, so $T_i = 2a/Rb = 2T_B$.

The virial coefficients must themselves be determined by the forces of interaction between the molecules because if there were no such forces all the virial

6.2 Intermolecular forces

coefficients would vanish and the gas would be ideal. In the next section we shall see what typical intermolecular forces are like, and then in §6.3 we shall see how, from the principles of statistical mechanics, one may calculate the second virial coefficient $B(T)$ when the intermolecular forces are given.

6.2 Intermolecular forces

In Fig. 6.1 is shown a typical potential energy of interaction $\phi(r)$ between non-bonding atoms or spherical or nearly spherical molecules, as a function of the distance r between their centers. It has qualitatively the same shape as the potential curve $V(r)$ of a diatomic molecule (as seen in Fig. 2.7, for instance) but is quantitatively different and has a different physical origin. The diatomic-molecule potential $V(r)$ is that associated with a real and typically strong chemical bond; its depth is a measure of the strength of the bond and the position of the minimum is the bonding distance. The depth ε of the minimum in $\phi(r)$ is typically less than that in $V(r)$ by a factor of 100 or more and the minimum in $\phi(r)$, at r_0, is much further out – perhaps twice as far or more – than that in $V(r)$.

An illustration of these quantitative differences between $\phi(r)$ and $V(r)$ is provided by the contrast between the $V(r)$ of diatomic chlorine and the interaction potential $\phi(r)$ of a pair of argon atoms. This is a good example to choose because the atomic numbers of argon and chlorine are so close – 18 and 17, respectively – so the neutral atoms have nearly equal numbers of electrons. The depth ε of the minimum in $\phi(r)$ for argon, as may be inferred from the heat of vaporization of the liquid or – by methods to be discussed in §6.3 – from the second

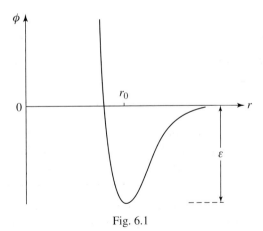

Fig. 6.1

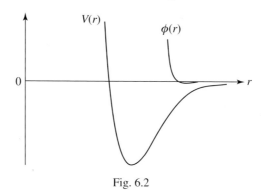

Fig. 6.2

virial coefficient $B(T)$, is about 1.0 kJ/mol, which is around 0.010 eV per atom pair; while the energy of dissociation of Cl_2 from its ground vibrational state is 2.5 eV, and the minimum in $V(r)$ is even deeper than that by an amount equal to the vibrational zero-point energy. [See the figure accompanying the solution to Exercise (3.1).] At the same time r_0 for argon–argon [again, as may be inferred from $B(T)$ by the methods to be discussed in §6.3] is about 3.9 Å, while the equilibrium internuclear distance in Cl_2 is 2.0 Å. These extreme quantitative differences between $V(r)$ and $\phi(r)$ are pictured schematically in Fig. 6.2. (Taking these to go to 0 at infinite separation is just a convention fixing the arbitrary zero of potential energy.)

The forces of which $\phi(r)$ is the potential are called van der Waals forces; they are the attractions and repulsions the van der Waals equation of state attempts to take into account via its parameters a and b. The force between centers is $-d\phi(r)/dr$, so for $r < r_0$, where $-d\phi(r)/dr > 0$, the force is one of repulsion (its being positive means that its direction is such as to tend to make r increase), while for $r > r_0$, where $-d\phi(r)/dr < 0$, it is a force of attraction.

The repulsion for $r < r_0$ is very strong; once $\phi(r)$, with decreasing r, crosses the r-axis (i.e., becomes positive), which it does at about $0.9r_0$, it increases extremely rapidly with further decrease of r. The origin of the sudden strong repulsion is partly the Coulomb repulsion between the electrons of the two atoms or molecules, and partly the Pauli exclusion principle, which would keep pairs of electrons (of the same spin) apart even in the absence of the Coulomb force. The repulsion sets in suddenly as the two electron clouds begin to interpenetrate and becomes rapidly stronger with increasing overlap (Fig. 6.3). This distance $\sigma(\simeq 0.9r_0)$ at which the strong repulsion suddenly sets in is a measure of the "size" of the atom or molecule; σ is twice its radius, hence its diameter. The repulsion is so strong and sets in so suddenly that the molecules may often be conveniently thought of as infinitely hard spheres – like billiard balls – of

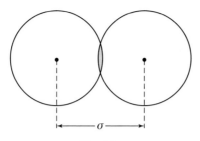

Fig. 6.3

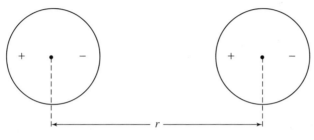

Fig. 6.4

diameter σ. It is the volume of this repulsive core of the intermolecular potential that is being measured by the van der Waals parameter b; the "volume" v_0 of the atom or molecule is $v_0 = (4/3)\pi(\sigma/2)^3 = (\pi/6)\sigma^3$, while, as we shall see (§6.3), $b = 4v_0$. (Here we use a molecular rather than molar van der Waals b; the b in Ex. (6.1) is the molar b, which exceeds the present one by a factor of Avogadro's number.)

The attractive component of the intermolecular potential has a different origin. In Fig. 6.4 are pictured two atoms a large distance r apart. They could be a pair of argon atoms in their (non-degenerate) ground states. Each one has, on average, a spherically symmetric distribution of electronic charge about its nucleus and so has no permanent dipole moment. Nevertheless, at any instant there are momentary imbalances in the electronic charge distributions, hence momentary dipole moments that are not 0 (even though they rapidly average to 0). Further, as long as the atoms are just close enough to feel each other's presence, there is a slight correlation between the direction of the instantaneous dipole in one atom with that in the other. This is seen in Fig. 6.4; at the instant that the dipole in the left-hand atom is oriented left-to-right, say ($+\!\!\!\rightarrow$), that in the right-hand atom is slightly more likely than not to be oriented in the same direction, for that produces an energetically favorable charge distribution in which the excess positive charge at one side of one of the atoms is closest to the excess negative charge on the other atom. The instantaneous dipole of one atom has thus polarized the other atom, the polarization then producing an attractive dipole–induced-dipole interaction. This polarization, we note, is only statistical: the (weak) correlation between the momentary charge imbalances on the two atoms produces configurations that are only slightly more often attractive than they are repulsive, resulting in a (weak) net attractive force. The correlations, hence the net attractions, are stronger at shorter distances, but become rapidly weaker as r increases and disappear at $r = \infty$. Thus, $-d\phi(r)/dr$ approaches 0 through a range of negative values as $r \to \infty$. With $\phi(r)$ itself approaching 0 as $r \to \infty$, by our choice for the 0 of potential energy, this means that $\phi(r)$, too, approaches its limit from the negative side, as in Figs. 6.1 and 6.2.

This explanation of the weak attractions between electrically neutral, non-bonding atoms or molecules at large separations is due to F. London. The attractive components of the van der Waals forces are therefore often called London forces – or sometimes dispersion forces, because London's quantitative treatment had some formal similarities to the theory of the electronic origins of optical dispersion.

The potential energy of dipole–dipole interactions is proportional to $1/r^3$, but since the London forces are only a correlation effect they appear only as a second-order perturbation, so the resulting $\phi(r)$ vanishes proportionally to $1/r^6$ at large r. (The same would be true if the molecules had permanent dipoles, but then the averaging would be thermal instead of electronic and the coefficient of $1/r^6$ would be inversely proportional to the absolute temperature T.) As the distance r increases and ultimately becomes comparable with the wavelength of a strong absorption line in the molecule's optical spectrum – 1000 Å or more, say – the $1/r^6$ potential goes over into a $1/r^7$ potential because of a relativistic effect called retardation. The attraction has already become so weak by then that the effect may almost always be ignored in the interactions between isolated pairs of atoms or molecules, but it may sometimes be important when it is integrated over many molecules, as in calculating the interaction between colloid particles or between pieces of bulk matter.

It is often convenient to have an analytical representation of the interaction energy $\phi(r)$ in Fig. 6.1 even if it is one that does not have much theoretical justification beyond its general shape. Of such analytical formulas the most popular is the *Lennard-Jones "6–12" potential*,

$$\phi(r) = \varepsilon[(r_0/r)^{12} - 2(r_0/r^6)] \tag{6.3}$$

with ε and r_0 as in Fig. 6.1. At large r, since the inverse sixth power vanishes more slowly than the inverse twelfth power, this $\phi(r)$ vanishes proportionally to $1/r^6$, and does so through negative values, which we know to be theoretically

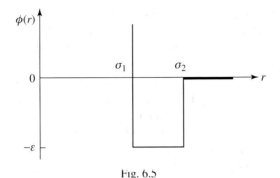

Fig. 6.5

justified. For small r, since the inverse twelfth power goes to ∞ more rapidly than the inverse sixth power, this $\phi(r)$ becomes infinite very rapidly – proportionally to $1/r^{12}$ – as $r \to \infty$. The rapid divergence of $\phi(r)$ as $r \to 0$ is realistic, but that it goes as an inverse twelfth power is purely empirical (and not always very accurate). We may easily verify by differentiation that the minimum in the $\phi(r)$ of (6.3) occurs at $r = r_0$, and we see that $\phi(r_0) = -\varepsilon$, again in accord with the figure. The curve of $\phi(r)$ vs. r crosses the r axis at $r = \sigma = 2^{-1/6} r_0 = 0.89 r_0$. This is the σ that is often taken to be the molecule's "diameter," as discussed earlier.

Another frequently used analytical approximation to $\phi(r)$ is the "square-well" potential pictured in Fig. 6.5 and defined by

$$\phi(r) = \begin{cases} \infty, & r < \sigma_1 \\ -\varepsilon, & \sigma_1 < r < \sigma_2 \\ 0, & r > \sigma_2. \end{cases} \quad (6.4)$$

This, too, corresponds to sudden, strong repulsion – now idealized as infinitely sudden and infinitely strong – at small distances, to attraction at intermediate distances, and to an interaction that falls off more-or-less rapidly – here idealized as vanishing identically – at larger distances. For many purposes even so highly idealized a representation of $\phi(r)$ as this one proves adequate. It is found to account well for the temperature dependence of the second virial coefficient $B(T)$ (see below, §6.3) when σ_2 is taken to be about $1.5\sigma_1$; i.e., when the width of the square well is about half the molecule's hard-sphere diameter, as in Fig. 6.5.

6.3 Second virial coefficient from statistical mechanics

We know that the virial coefficients $B(T), C(T), \ldots$ must somehow be expressible in terms of the intermolecular interaction potential $\phi(r)$ because this

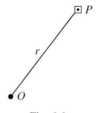

Fig. 6.6

is what gives rise to them; if $\phi(r)$ were 0 at all distances r the gas would be ideal and the virial coefficients would be 0 at all temperatures T. The object of this section is to derive from statistical mechanics an explicit formula showing how the second virial coefficient $B(T)$ follows from $\phi(r)$, and then to illustrate the applications of that formula.

Suppose we know that there is a molecule at some point O in the volume, what is then the average density at any point such as P, distant r from O? (Fig. 6.6.) If $\rho = N/V$ is the overall average number density in the gas then the mean density at P *given* that there is a molecule at O would differ from ρ by some factor – call it $g(r)$ – that takes account of the correlations in the distribution of molecules arising from the forces they exert on each other:

$$\text{(mean local density at distance } r \text{ from } O) = \rho g(r). \qquad (6.5)$$

This function $g(r)$ is called the radial distribution function, or pair distribution function, and plays an important role in the theory of the liquid state (Chapter 7) as well as in that of the non-ideal gas.

As long as the gas is dilute the correlations in the positions of the molecules that $g(r)$ takes account of are due to the potential energy $\phi(r)$ that a molecule at P (Fig. 6.6) feels owing to the presence of the molecule at O. By the Boltzmann distribution law (§1.2) we then have simply

$$g(r) = e^{-\phi(r)/kT}. \qquad (6.6)$$

[We have here accepted the convention $\phi(\infty) = 0$, as in Figs. 6.1 and 6.2. Otherwise, we would have written (6.6) with $\phi(r) - \phi(\infty)$ in place of $\phi(r)$.] If $\phi(r)$ were just 0 at all r – i.e., if the molecules did not exert forces on each other – we would have $g(r) = 1$ at all r. Then from (6.5) the mean local density at P would be just the overall mean density ρ; the presence of the molecule at O would exert no influence on the presence or absence of any other molecule at any distance from O, and the gas would be ideal. As long as there is a $\phi(r)$ the attractions and repulsions between molecules make the mean local density different from the overall ρ. At large enough distances, however, the forces

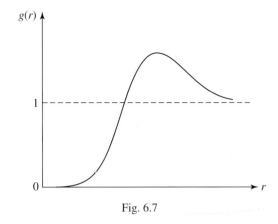

Fig. 6.7

between molecules are so weak that these correlations are lost. That, too, is contained in (6.6), which tells us that $g(\infty) = 1$.

With $\phi(r)$ as in Fig. 6.1, the $g(r)$ from (6.6) is as pictured in Fig. 6.7. Because of the strong repulsion between molecules at close approach it is very unlikely that another molecule will be found very close to the given molecule at O. The mean local density at small r must then be much less than the overall ρ; i.e., $g(r)$ must be much less than 1 at small r, as seen in the figure. Molecules attract each other at intermediate distances r, so a larger than average number of molecules will be found at such distances from the molecule at O; that is, at such distances $g(r)$ must exceed 1, again as it is seen to do in Fig. 6.7. Finally, at large r, as we already remarked, $g(r)$ goes to 1. Since $\phi(r)$ approaches 0 through negative values as $r \to \infty$, correspondingly, from (6.6), $g(r)$ approaches 1 from a range of values greater than 1, as also seen in the figure.

We said that the formula (6.6) for $g(r)$ would hold as long as the gas was dilute, for then the only correlating effect at P that a molecule at O has (Fig. 6.6) is through the direct action of the potential $\phi(r)$. If the gas were not so dilute $g(r)$ would be more complicated, because it would have to take account of the fact that the molecules attracted to or repelled by the molecule at O also attract and repel each other. The result is that (6.6) is only a low-density approximation to the true $g(r)$, which may sometimes need to be corrected. The correction terms, like those in the virial series (6.2), could be expressed as a series in increasing powers of the density $\rho = N/V$, each term being a successively refined correction to the preceding. Thus,

$$g(r) = e^{-\phi(r)/kT} + \rho g_1(r) + \rho^2 g_2(r) + \cdots, \tag{6.7}$$

in which additional functions $g_1(r)$, $g_2(r)$, etc. appear with each new power of ρ. These g_1, g_2, etc. may depend on the temperature T as well as being functions of r, but they do not depend on ρ, which occurs in (6.7) only in the powers ρ, ρ^2, etc. For purposes of calculating the second virial coefficient $B(T)$, however, these corrections, as we shall see, are not needed; the low-density approximation (6.6) to $g(r)$ is sufficient. Had we wished also to calculate the third virial coefficient $C(T)$ we would have had to incorporate the first correction $\rho g_1(r)$ in $g(r)$; for the fourth virial coefficient $D(T)$ we would have needed both $\rho g_1(r)$ and $\rho^2 g_2(r)$; etc. With each additional correction term the resulting $g(r)$ becomes more complex, and develops a series of maxima and minima. Even in a liquid, though, as we shall see in Chapter 7, some features of the simple $g(r)$ in Fig. 6.7 persist: $g(r)$ remains vanishingly small for small r, has a prominent peak at intermediate r (although then due more to the repulsive than to the attractive forces, and followed by a succession of subsidiary peaks and valleys), and ultimately approaches 1 as $r \to \infty$.

The total energy E of the gas is the sum $E_{\text{i.g.}} + E_{\text{intermol}}$ of the energy $E_{\text{i.g.}}$ the gas would have had if there were no intermolecular forces acting – i.e., if the gas were ideal – and E_{intermol}, the potential energy of intermolecular interaction. In Chapter 2 we learned how to calculate $E_{\text{i.g.}}$: for N molecules there is a contribution $(3/2)NkT$ from the translational degrees of freedom (motion of the molecules' centers of mass), with additional contributions from the molecules' rotational and vibrational degrees of freedom if there are such. But however complex the molecules – whatever their internal structure – this $E_{\text{i.g.}}$ is a function of temperature alone, not of density. Any density dependence of the total energy E comes entirely from the second term, E_{intermol}.

Since the local density at P distant r from a molecule at O (Fig. 6.6) is $\rho g(r)$, the infinitesimal "number of molecules" in an infinitesimal volume $d\tau$ centered at P (really, the infinitesimal probability of finding a molecule in $d\tau$, times the total number N of such molecules), is $\rho g(r) d\tau$. The potential energy of interaction of any molecule in that volume element $d\tau$ at P, with the molecule at O, is $\phi(r)$, so the total potential energy of interaction between the molecule at O and those in $d\tau$ is $\rho g(r) \phi(r) d\tau$. The sum total of the energies of interaction of the molecule at O with all the other molecules of the gas is then the sum (integral) of this over all such volume elements $d\tau$ at all such points P in space:

$$\rho \int g(r) \phi(r) d\tau. \tag{6.8}$$

If we multiply this by N, the total number of molecules (each one in turn playing the role of the central molecule at O),

$$N\rho \int g(r) \phi(r) d\tau, \tag{6.9}$$

6.3 Second virial coefficient

we might, at first, suppose that we would then have the total E_{intermol}. What (6.9) actually is, though, is *twice* E_{intermol}. That is because (6.9) counts every interaction twice: it counts the interaction between molecules i and j when i is the central molecule and j is one of the molecules in the rest of the gas with which i interacts, and it counts that same interaction again when it is j that is the central molecule and i is one of the others with which *it* interacts. Thus, finally,

$$E_{\text{intermol}} = \frac{1}{2} N \rho \int g(r)\phi(r) \mathrm{d}\tau. \tag{6.10}$$

We have been supposing, for simplicity, that the molecules are spherical or nearly so, so that the interaction between them is describable by a spherically symmetric $\phi(r)$. If the three-dimensional integration in (6.10) is then expressed in polar coordinates, for which the element of volume $\mathrm{d}\tau$ is $r^2 \sin\theta \mathrm{d}\phi \mathrm{d}\theta \mathrm{d}r$ with θ and ϕ the polar and azimuthal angles, the integration over the angles may be done immediately (over θ from 0 to π and over ϕ from 0 to 2π), producing a factor 4π and leaving an integral over the radial coordinate r alone:

$$E_{\text{intermol}} = 2\pi N\rho \int_0^\infty g(r)\phi(r) r^2 \mathrm{d}r. \tag{6.11}$$

To be sure, the volume V of the gas is finite, so any element of volume $\mathrm{d}\tau$ in V is necessarily at a finite distance r from any origin O in V, but because V is *macro*scopically large while the interaction potential $\phi(r)$ effectively becomes 0 once r is large on a *micro*scopic scale, the integration in (6.11) can go out to $r = \infty$ with no real error. Equation (6.11) is really a formula for E_{intermol}/N in the thermodynamic limit in which N and V both go to ∞ at fixed $\rho = N/V$.

We now make use of the expression for $g(r)$ in (6.6) or (6.7), and thus find for the total interaction energy per molecule

$$E_{\text{intermol}}/N = 2\pi\rho \int_0^\infty e^{-\phi(r)/kT} \phi(r) r^2 \mathrm{d}r + \cdots, \tag{6.12}$$

where "$+\cdots$" represents correction terms proportional to ρ^2, ρ^3, etc. To leading order in ρ, as we see, E_{intermol}/N is proportional to ρ. In the limit of vanishing density, $\rho \to 0$, the contribution of E_{intermol} to the total energy E thus vanishes and E becomes just $E_{\text{i.g.}}$: the gas is ideal.

Think of N as fixed, so that varying density means varying V. Then for the rate of change of the total energy E with density or volume we have

$$\left(\frac{\partial E}{\partial V}\right)_T = \left[\frac{\partial (E/N)}{\partial (1/\rho)}\right]_T = -\rho^2 \left[\frac{\partial (E/N)}{\partial \rho}\right]_T \tag{6.13}$$

(because $d\rho^{-1} = -\rho^{-2}d\rho$). But since $E_{i.g.}$ is a function of temperature alone the only source of density dependence in $E = E_{i.g.} + E_{intermol}$ is from the intermolecular interaction term. Therefore (6.13) becomes

$$\left(\frac{\partial E}{\partial V}\right)_T = -\rho^2 \left[\frac{\partial (E_{intermol}/N)}{\partial \rho}\right]_T, \qquad (6.14)$$

or, from (6.12),

$$\left(\frac{\partial E}{\partial V}\right)_T = -2\pi\rho^2 \int_0^\infty e^{-\phi(r)/kT} \phi(r) r^2 dr + \cdots \qquad (6.15)$$

with correction terms of order ρ^3, ρ^4, etc.

The thermodynamic identity $(\partial E/\partial V)_T = -p + T(\partial p/\partial T)_V$ (now using E for energy in place of U) may be rewritten

$$(\partial E/\partial V)_T = -[\partial (p/kT)/\partial (1/kT)]_\rho. \qquad (6.16)$$

(With N fixed, fixed $\rho = N/V$ is the same as fixed V; and $dT^{-1} = -T^{-2}dT$.) Thus, the left-hand side of (6.15) is the derivative of $-p/kT$ with respect to $1/kT$. But the integral on the right-hand side of (6.15) is also a derivative with respect to $1/kT$:

$$\int_0^\infty e^{-\phi(r)/kT} \phi(r) r^2 dr = \frac{d}{d\frac{1}{kT}} \int_0^\infty \left(1 - e^{-\phi(r)/kT}\right) r^2 dr. \qquad (6.17)$$

This can be verified by carrying out the indicated differentiation. The constant 1 inside the parentheses in the integrand on the right-hand side could have been any constant as far as the differentiation with respect to $1/kT$ is concerned, but it is the constant 1 that secures the convergence of the integral, because (by the convention we have been using) $\phi(\infty) = 0$, so with the constant 1 the integrand goes to 0 as the integration variable goes to ∞.

Both sides of (6.15) have now been expressed as derivatives with respect to $1/kT$; we have

$$-\left(\frac{\partial \frac{p}{kT}}{\partial \frac{1}{kT}}\right)_\rho = -2\pi\rho^2 \frac{d}{d\frac{1}{kT}} \int_0^\infty \left(1 - e^{-\phi(r)/kT}\right) r^2 dr + \cdots \qquad (6.18)$$

with correction terms of order ρ^3, ρ^4, etc. Therefore

$$p/kT = h(\rho) + 2\pi\rho^2 \int_0^\infty \left(1 - e^{-\phi(r)/kT}\right) r^2 dr$$
$$+ \text{corrections of order } \rho^3, \rho^4, \ldots, \qquad (6.19)$$

6.3 Second virial coefficient

where $h(\rho)$ is some function of ρ alone, independent of the temperature (because it must disappear when differentiated with respect to $1/kT$ at fixed ρ), which we have to determine. But as $\rho \to 0$ at fixed T we must retrieve the ideal-gas law, $p = NkT/V = \rho kT$; so the function $h(\rho)$ can be nothing other than ρ itself. Therefore,

$$p/kT = \rho \left[1 + 2\pi \rho \int_0^\infty \left(1 - e^{-\phi(r)/kT}\right) r^2 dr \right.$$
$$\left. + \text{corrections of order } \rho^2, \rho^3, \ldots \right]. \qquad (6.20)$$

But this (cf. (6.2)) is the virial series, since $\rho = N/V$. We have therefore explicitly calculated the second virial coefficient $B(T)$ in terms of the potential $\phi(r)$ of the intermolecular forces:

$$B(T) = 2\pi \int_0^\infty \left(1 - e^{-\phi(r)/kT}\right) r^2 dr. \qquad (6.21)$$

Exercise (6.2). Calculate $B(T)$ for the square-well potential in Fig. (6.5).

Solution. Break up the range of integration in (6.21) into the three separate ranges $0 < r < \sigma_1$, $\sigma_1 < r < \sigma_2$, and $\sigma_2 < r < \infty$. By (6.4), $\phi(r)$ is $+\infty$ in the first range, where the integrand is then just r^2; $\phi(r)$ is the constant $-\varepsilon$ in the second range, where the integrand is then just $[1 - \exp(\varepsilon/kT)]r^2$; and $\phi(r)$ is identically 0 in the third range, where the integrand is then 0 and contributes nothing to the integral. Thus,

$$B(T) = 2\pi \left[\int_0^{\sigma_1} r^2 \, dr + \left(1 - e^{\varepsilon/kT}\right) \int_{\sigma_1}^{\sigma_2} r^2 dr \right]$$
$$= (2\pi/3)\left[\sigma_1^3 - \left(e^{\varepsilon/kT} - 1\right)\left(\sigma_2^3 - \sigma_1^3\right)\right].$$

Exercise (6.3). We are given an intermolecular interaction potential $\phi(r)$ of the form

$$\phi(r) = -\varepsilon \ln\left(1 + Ae^{-r/a}\right)$$

with three parameters, ε, a, and A, of which the first two are positive, while the third may be positive or negative but is greater than -1.

(a) Sketch $\phi(r)$ as a function of r when A is positive and again when A is negative, and note that $\phi(r)$ is an entirely attractive potential in the first case and entirely repulsive in the second.

(b) Show that at the temperature $T = \varepsilon/k$ (k = Boltzmann's constant) the second virial coefficient B of a gas of molecules interacting with this potential is $-4\pi Aa^3$.

Solution.

(a)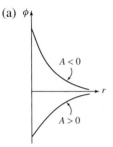

We see from the sketch that the sign of $-\mathrm{d}\phi(r)/\mathrm{d}r$ is opposite that of A for all r. Thus, $\phi(r)$ is entirely attractive for $A > 0$ and entirely repulsive for $A < 0$.

(b) When $T = \varepsilon/k$,

$$\mathrm{e}^{-\phi(r)/kT} = \mathrm{e}^{\ln(1+A\mathrm{e}^{-r/a})} = 1 + A\mathrm{e}^{-r/a}.$$

Then

$$B = 2\pi \int_0^\infty \left(1 - \mathrm{e}^{-\phi(r)/kT}\right) r^2 \mathrm{d}r$$
$$= -2\pi A \int_0^\infty \mathrm{e}^{-r/a} r^2 \mathrm{d}r = -2\pi Aa^3 \int_0^\infty \mathrm{e}^{-x} x^2 \mathrm{d}x.$$

The last integral has the value 2, so

$$B = -4\pi Aa^3,$$

as was to be shown. Note that B is positive if the potential is repulsive ($A < 0$) and negative if the potential is attractive ($A > 0$), as expected.

At high temperatures, where the thermal energy kT is much greater than the depth ε of the attractive well in the potential in Ex. (6.2), so that $\mathrm{e}^{\varepsilon/kT} \simeq 1 + \varepsilon/kT$, the second virial coefficient calculated there becomes

$$B(T) \simeq (2\pi/3)\left[\sigma_1^3 - \left(\sigma_2^3 - \sigma_1^3\right)\varepsilon/kT\right]. \quad (6.22)$$

On comparing this with the result of Ex. (6.1) in §6.1 we arrive at an identification of the van der Waals constants a and b in terms of the parameters

ε, σ_1, and σ_2 of the square-well potential in Fig. 6.5. As in the reference to the van der Waals constant b in §6.2, we shall again here use b to mean the molecular b, related to the molar b by a factor of Avogadro's number N_0, and shall likewise use a to mean the molecular a, related to the molar a by a factor of N_0^2. In these terms the second virial coefficient $B(T)$ for the van der Waals gas found in Ex. (6.1) is

$$B(T) = b - a/kT \tag{6.23}$$

(note k in place of R); whereupon we make the identifications

$$b = (2\pi/3)\sigma_1^3, \qquad a = (2\pi/3)(\sigma_2^3 - \sigma_1^3)\varepsilon. \tag{6.24}$$

The σ_1 of the square-well potential is the hard-sphere diameter σ depicted in Fig. 6.3, in terms of which we earlier (§6.2) expressed the "volume" v_0 of the atom or molecule: $v_0 = (\pi/6)\sigma^3$. We therefore have the identification $b = 4v_0$: the van der Waals b is four times the volume of the molecule. At the same time $(4\pi/3)(\sigma_2^3 - \sigma_1^3)$ is the volume, call it v_1, of the attractive mantle of the intermolecular potential, which is seen in Fig. 6.5 to surround the repulsive core. In terms of that volume and of the depth ε of the potential in the attractive mantle we may now, from (6.24), identify the van der Waals a as $\frac{1}{2}v_1\varepsilon$.

This calculation illustrates an important application of the formula (6.21) for $B(T)$: with that formula, experimental measurement of $B(T)$ as a function of temperature yields quantitative information about the intermolecular forces. If we imagine, hypothetically, that $B(T)$ is found by experiment actually to be of the form given by (6.23), $B(T) = b - a/kT$ with two constants a and b, and if we suppose the intermolecular potential $\phi(r)$ to be adequately represented by the square well in Fig. 6.5, then, as we just saw, the experimental a and b would tell us the values of σ_1 and $(\sigma_2^3 - \sigma_1^3)\varepsilon$. In reality the experimental $B(T)$ is not very well fit by $b - a/kT$ with only two adjustable constants a and b; it is fit much better by the form derived in Ex. (6.2) before we made the subsequent approximation $\exp(\varepsilon/kT) - 1 \simeq \varepsilon/kT$; viz., by the form

$$B(T) = (2\pi/3)\left[\sigma_1^3 - \left(e^{\varepsilon/kT} - 1\right)(\sigma_2^3 - \sigma_1^3)\right] \tag{6.25}$$

with the three adjustable constants σ_1, σ_2, and ε. It is no surprise that a formula with three adjustable parameters can be made to fit better than one with only two, but in this instance the improvement is largely because of the more realistic functional form at low temperatures: $B(T)$ goes rapidly to large negative values proportionally to $-\exp(\varepsilon/kT)$ rather than as slowly as $-a/kT$.

The function $B(T)$ in (6.25), which in its form is nearly indistinguishable from what is found experimentally, is sketched in Fig. 6.8. On fitting

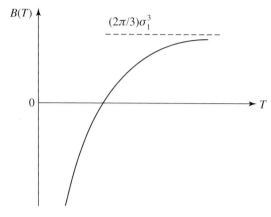

Fig. 6.8

this to experiment one finds a best fit with σ_2 about $1.5\sigma_1$, as anticipated in Fig. 6.5.

The Lennard-Jones 6–12 potential (6.3) is more realistic than the square-well potential; it looks like the $\phi(r)$ in Fig. 6.1. Although the 6–12 potential has only two adjustable parameters, r_0 and ε, rather than three, the $B(T)$ that is found by (numerical) evaluation of the integral in (6.21) with that $\phi(r)$, with r_0 and ε chosen to give the best fit, fits the experimental $B(T)$ very well. That has been the source of much of our knowledge of intermolecular forces; the ε and r_0 found from the fitting tell us the depth of the attractive well in the intermolecular potential and, through $\sigma \simeq 0.9 r_0$, the molecule's diameter. (Other important sources of such information are from measurement of the transport properties – viscosity, diffusion coefficient, and thermal conductivity – in dilute gases and from the measurement of scattering cross-sections in molecular beams.)

Even if we had no preconceived notion of what its shape was we would immediately learn something important about $\phi(r)$ just from the fact that experimentally $B(T)$ is positive for some T and negative for other T (cf. Fig. 6.8). From the formula (6.21) that fact alone tells us that $\phi(r)$ must be positive for some r and negative for other r; for if $\phi(r)$ were positive, say, for all r, then the integrand in (6.21) would also be positive for all r and that would make $B(T)$ positive for all T, while if $\phi(r)$ were negative for all r the integrand in (6.21) would also be negative for all r and that would make $B(T)$ negative for all T.

Once the second virial coefficient $B(T)$ has been determined, whether by experiment or by theory, it may be used to calculate non-ideality corrections to any of the thermodynamic functions of a dilute gas, using known thermodynamic

6.3 Second virial coefficient

identities to relate those functions to the pressure. For example, from the identity (6.16) we referred to earlier, and the virial series (6.2), we have

$$\left(\frac{\partial E}{\partial V}\right)_T \simeq kT^2 \left(\frac{N}{V}\right)^2 \frac{dB(T)}{dT} \qquad (6.26)$$

(where we have again used $dT^{-1} = -T^{-2}dT$). We have here neglected contributions from the third and higher virial coefficients, which is permissible as long as the density is not very great. Now, we know E to be $E_{\text{i.g.}} + E_{\text{intermol}}$, where E_{intermol}, the correction for non-ideality, vanishes as $N/V \to 0$ (see above); while $E_{\text{i.g.}}$ is independent of N/V – it depends on temperature alone. Therefore, from (6.26),

$$E \simeq E_{\text{i.g.}} - kT^2 \frac{N^2}{V} \frac{dB(T)}{dT}, \qquad (6.27)$$

which expresses the correction E_{intermol} to $E = E_{\text{i.g.}}$ in terms of $B(T)$.

Similarly, from the thermodynamic identity $(\partial S/\partial V)_T = (\partial p/\partial T)_V$ we have

$$\left(\frac{\partial S}{\partial V}\right)_T \simeq \frac{Nk}{V}\left[1 + B(T)\frac{N}{V}\right] + kT\left(\frac{N}{V}\right)^2 \frac{dB(T)}{dT}$$

$$= \frac{Nk}{V} + k\left(\frac{N}{V}\right)^2 \left[B(T) + T\frac{dB(T)}{dT}\right]. \qquad (6.28)$$

This implies

$$S \simeq Nk \ln \frac{V}{Nv(T)} - k\frac{N^2}{V}\left[B(T) + T\frac{dB(T)}{dT}\right], \qquad (6.29)$$

where $v(T)$ is some function of T alone, independent of V (but with the dimensions of a volume). The first term, $Nk \ln[(V/N)/v(T)]$, is $S_{\text{i.g.}}$, the entropy of the ideal gas: $Nk \ln(V/N)$ is as in (2.25), while $v(T)$ consists of the remaining contributions from the translational degrees of freedom of the molecules, as given in (2.25), together with all the contributions from the internal degrees of freedom. It is the terms in B and dB/dT in (6.29) that are the corrections for non-ideality. Thus,

$$S - S_{\text{i.g.}} \simeq -k\frac{N^2}{V}\left[B(T) + T\frac{dB(T)}{dT}\right]. \qquad (6.30)$$

This is $S(V, T) - S_{\text{i.g.}}(V, T)$. We have also $S(p, T) - S_{\text{i.g.}}(p, T) = -Np\, dB(T)/dT$. It makes a difference, when we turn off the interactions to produce the ideal gas, whether we do so at fixed V (in which case p changes) or at fixed p (in which case V changes). These formulas relating $S - S_{\text{i.g.}}$ to $B(T)$ are what allow one to correct for non-ideality of the gas when comparing theoretical and experimental entropies in connection with the third law of thermodynamics (Chapter 5, §5.3).

6.4 Long-range forces

In the course of the derivation of the formula (6.21) for $B(T)$ in terms of the intermolecular interaction potential $\phi(r)$ it was remarked that only when the constant that appeared inside the parentheses in the integrand was 1 could the integral converge. That is because $\phi(\infty) = 0$, so when that constant is 1, and only then, the integrand approaches 0 as $r \to \infty$; the integral would certainly diverge otherwise. The conclusion that the constant that is added to $-\exp[-\phi(r)/kT]$ in the integrand must be 1 is not dependent on the arbitrary convention $\phi(\infty) = 0$, for with any other choice of $\phi(\infty)$ it would have been not $\phi(r)$ but $\phi(r) - \phi(\infty)$ that appeared in the integrand, and that would again have approached 0 as r became infinite.

We thus see that it is necessary for the convergence of the integral that the constant be 1; but it is not sufficient – the integral may still diverge. That happens when $\phi(r)$ (or $\phi(r) - \phi(\infty)$) approaches 0 too slowly as $r \to \infty$. As $\phi(r)$ approaches 0 at large r the exponential $\exp[-\phi(r)/kT]$ is more and more closely approximated by $1 - \phi(r)/kT$, so in integrating out to $r = \infty$ in the formula (6.21) we find ourselves integrating

$$\int^{\infty} \phi(r) r^2 \mathrm{d}r. \tag{6.31}$$

If $\phi(r)$ vanishes more rapidly than $1/r^3$ as $r \to \infty$ (for example, if it vanishes as $1/r^6$, which we saw in §6.2 to be characteristic of the attractive London dispersion forces) then the integrand vanishes more rapidly than $1/r$, this contribution to the total integral is finite, and there is no problem: everything is as we supposed in §6.3. But if $\phi(r)$ vanishes as slowly as $1/r^3$, the integral (6.31) becomes $\ln r$ evaluated at $r = \infty$, which is infinite; and if $\phi(r)$ vanishes even more slowly than $1/r^3$ the divergence of the integral is even more rapid. That is so, for example, with Coulomb forces, for which $\phi(r) \sim 1/r$. A force derived from an intermolecular potential that vanishes as slowly as $1/r^3$ or more slowly as $r \to \infty$ (so, a force $-\mathrm{d}\phi/\mathrm{d}r$ that vanishes as slowly as $1/r^4$ or more slowly) is said to be long-ranged. When the forces are short-ranged the virial coefficient B is finite; when they are long-ranged, B is infinite.

6.4 Long-range forces

What does it mean when the virial coefficient is infinite? Here is a simple mathematical example. We expand a function $f(x)$ in powers of x,

$$f(x) = f(0) + \frac{1}{1!}f'(0)x + \frac{1}{2!}f''(0)x^2 + \cdots + \frac{1}{n!}f^{(n)}(0)x^n + \cdots, \tag{6.32}$$

where $f'(0), f''(0), \ldots, f^{(n)}(0), \ldots$ are the 1st, 2nd, ..., n^{th}, ... derivatives of $f(x)$ evaluated at $x = 0$. It is an expansion "about" $x = 0$; the leading term $f(0)$ is just $f(x)$ evaluated at $x = 0$, and, when x is small, the succeeding terms are successively smaller corrections. If x were N/V and f were some thermodynamic function this would be a virial series. With $f(x) = 1/(1-x)$ we may easily verify that $f^{(n)}(0) = n!$, so that (6.32) becomes the familiar expansion $1/(1-x) = 1 + x + x^2 + x^3 + \cdots$. But what if we apply the formula (6.32) to the function $f(x) = 1/(1-\sqrt{x})$? The leading term is still $f(0) = 1$, but now the coefficient $f'(0)$ of the first power of x is ∞: $f'(x) = 1/[2\sqrt{x}(1-\sqrt{x})^2]$, which is ∞ at $x = 0$. The reason that coefficient is infinite is that, for the function $1/(1-\sqrt{x})$, the expansion (6.32) is of the wrong form. The correct expansion of $1/(1-\sqrt{x})$ for small x would not be in integer powers of x but in powers of $\sqrt{x}$: obviously $1/(1-\sqrt{x}) = 1 + \sqrt{x} + (\sqrt{x})^2 + \cdots$. The true first correction to the leading term 1 is then not of order x but is $\sqrt{x}$, which for small x is very much greater than x. The coefficient of x was infinite when we tried to expand in integer powers of x because it was trying to make up for the x being infinitely too small; that first correction term was supposed to be $\sqrt{x}$, which, as $x \to 0$, is infinitely greater than x. ($\sqrt{x}/x = 1/\sqrt{x} \to \infty$ as $x \to 0$.)

The virial coefficient B is infinite when the forces are of long range because the correct expansion of p/kT at low density is then no longer in integer powers of N/V. The first correction to the ideal-gas law $pV/NkT = 1$ is then not of the form BN/V but, if it is a power of N/V, then it is a fractional power, between 0 and 1. The divergence of B was telling us that N/V is infinitely too small; the real first correction is something that in the limit $N/V \to 0$ is infinitely greater than N/V.

That is what happens with Coulomb forces. When there are equal numbers of positive and negative charges, so overall charge neutrality, the leading correction to the ideal-gas law is of order $(N/V)^{1/2}$; thus, at low densities,

$$p/kT = (N/V)[1 + B'(N/V)^{1/2} + \cdots] \tag{6.33}$$

with some coefficient B' that is a function of temperature alone. A "plasma" consisting of positive ions or bare nuclei plus electrons, such as one would have at the extraordinarily high temperatures produced in nuclear fusion or in stellar interiors, or even at low temperatures but high dilution in the earth's ionosphere and magnetosphere, would be an example. That the leading correction to the ideal-gas law is then of order $(N/V)^{1/2}$ instead of N/V is something we really already know from the Debye-Hückel theory of electrolyte solutions. We learn there that an ionic activity coefficient deviates from unity by an amount that, at great dilution, is proportional to the square-root of the ionic strength. The latter is just a (charge-weighted) measure of the ionic concentration or number density, N/V. That activity coefficient then corrects the ideal osmotic-pressure law $\pi V = NkT$. The correction is exactly as in (6.33). The osmotic pressure π of a liquid solution is entirely analogous to the pressure p of a gas; deviations from ideal-dilute solution behavior are due to solute–solute interactions (as mediated by the solvent) just as deviations from ideal-gas behavior are due to the interactions between the molecules in the gas.

7
The liquid state

7.1 Structure of liquids

In one of the twentieth century's greatest didactic works on science, the encyclopedic *Course of Theoretical Physics* of Landau and Lifshitz, there appears the following statement:

Unlike solids and gases, liquids do not allow a general calculation of their thermodynamic quantities or even their temperature dependence. The reason for this is the presence of strong interactions between the molecules of the liquid without having at the same time the smallness of the vibrations which makes the thermal motion of solids so simple. The high intensity of the molecular interaction makes it important to know, when calculating thermodynamic quantities, the actual law of interaction, which varies for different liquids. The only thing which can be done in general form is the study of the properties of liquids near absolute zero. The principles involved in this question are of considerable interest although in practice there exists only one substance (helium) which can remain liquid down to absolute zero.*

At the time that was written it was indeed true that there was no general theory of liquids – that the only liquid whose properties could be derived from statistical mechanics was liquid helium, and then only in the neighborhood of the absolute zero of temperature. Now, a generation later, the situation has been wholly transformed, and we are able to calculate the properties of ordinary liquids with nearly as much assurance as we do those of dilute gases and harmonic solids. It is true, as remarked in the quoted passage, that the central difficulty in a theory of liquids is that the system cannot be decomposed into independent or only weakly interacting degrees of freedom, like the nearly independent molecules of a dilute gas or the nearly independent modes of vibration of a rigid solid. One must somehow deal with the formidable complexity of a system of strongly

---

* L.D. Landau and E.M. Lifshitz, *Course of Theoretical Physics 5: Statistical Physics*, transl. by E. Peierls and R.F. Peierls (Pergamon Press, 1958), Chap. VI, §66, pp. 198–199.

interacting degrees of freedom. What, then, led to the breakthrough, and what is the nature of the new theory?

The physical idea that underlies the modern theory is not itself new but goes back to van der Waals in the nineteenth century.* What is new is our ability to realize van der Waals's vision through the intervention of high-speed digital computing. The basic idea is that the structure of a liquid – the spatial arrangements of its molecules relative to each other – is determined mostly by the sudden strong repulsions between molecules when they come close together (§6.2), and that once the structure is thus determined the effects of the much weaker and more slowly varying attractive forces can be adequately treated by a simple approximation.

We remarked in §6.2 that the strong repulsions between molecules can often be idealized as those between hard spheres. Then if it is true that the structure of a liquid is determined primarily by those repulsions, the structure of a real liquid (of spherical or nearly spherical molecules) should be nearly the same as that of a hypothetical fluid of non-attracting hard spheres of diameter equal to the real molecule's σ (§6.2 and Fig. 6.3) at the same number density. The relevant measure of *structure* here is the radial distribution function (pair distribution function) $g(r)$ introduced in §6.3. This, it is to be recalled, is the quantity by which the overall average density ρ is to be multiplied to obtain the mean local density at the distance r from any given molecule. If the energy of interaction among molecules is additive, so that the total is just the sum of the interactions of all pairs, then this total interaction energy, and so ultimately all of the liquid's thermodynamic properties including its equation state, is given in terms of $g(r)$ by (6.10) or (6.11). (Intermolecular forces are in reality not strictly additive; the energy of interaction of a triple of molecules, for example, is not exactly the sum of the interactions of the three pairs taken separately. Chemical forces, characterized by the saturation of valency, are an extreme example of non-additivity. But with ordinary, non-chemical, non-specific van der Waals forces such deviations from additivity, while measurable, are not great; they affect the total interaction energy in a liquid by only about 10 or 15%, typically.)

We wish, therefore, to compare the $g(r)$ of a liquid of simple molecules with that of a fluid of hard spheres of comparable size at a comparable density. In Fig. 7.1 is the experimentally determined radial distribution function of liquid argon when it is in equilibrium with its vapor at 85 K, which is just slightly above the triple-point temperature, 84 K. It was obtained from measurements of the angular dissymmetry of the scattering of slow neutrons from ^{36}Ar. The

---

* J.S. Rowlinson, *Van der Waals and the physics of liquids*, introductory essay in *J.D. van der Waals: On the Continuity of the Gaseous and Liquid States* (Studies in Statistical Mechanics XIV), J.S. Rowlinson, ed. (North-Holland, 1988).

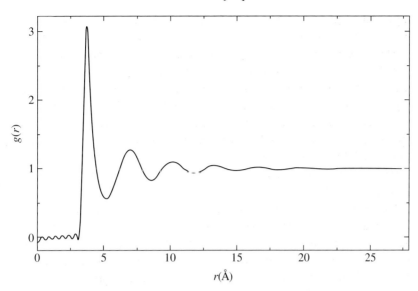

Fig. 7.1 From J.L. Yarnell, M.J. Katz, R.G. Wenzel, and S.H. Koenig, *Phys. Rev.* A **7** (1973) 2130. Reproduced with permission.

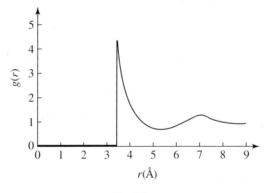

Fig. 7.2

oscillations one sees between $r = 0$ and $r = 3.4$ Å, which is the diameter σ for argon, are an artifact of numerical Fourier inversion; the directly measured quantity was not $g(r)$ itself but the Fourier transform of $g(r) - 1$. In Fig. 7.2 is $g(r)$ for a fluid of hard spheres of diameter $\sigma = 3.4$ Å, at the same number density, 2.1×10^{22} molecules/cm^3, as in the argon measurements. The structure of a hard-sphere fluid, and so in particular the hard-sphere $g(r)$, is independent of temperature; the temperature determines only how fast the spheres move but not their average locations. This hard-sphere $g(r)$ was obtained by computer simulation, by methods to be described later (§§7.3 and 7.4).

While there are discernible differences between the two figures, the similarities (taking account of the difference in scale: Fig. 7.1 covers a much greater range of r) are striking. In Fig. 7.1, $g(r)$ is practically, and in Fig. 7.2 it is exactly, 0 out to $r = \sigma = 3.4$ Å, where it then jumps almost discontinuously (Fig. 7.1) or actually discontinuously (Fig. 7.2) to a high value, 3.0 in Fig. 7.1 and 4.4 in Fig. 7.2. It then goes through a local minimum value of 0.6 near $r = 5$ Å (Fig. 7.1), to be compared with a local minimum value of 0.7 at $r = 5.3$ Å (Fig. 7.2); then a local maximum of 1.25 near $r = 7$ Å (Fig. 7.1), to be compared with a maximum of 1.27 at 6.9 Å (Fig. 7.2); and at the end of the range of r in Fig. 7.2 there are also the beginnings of the next minimum.

The oscillations in $g(r)$ seen in Figs. 7.1 and 7.2 for $r > \sigma$ are as we anticipated in §6.3 for a dense fluid. This contrasts with the $g(r)$ of a dilute gas earlier depicted in Fig. 6.7. Further, since the high peak in the experimental $g(r)$ near $r = 3.4$ Å correlates so well with that for non-attracting hard spheres it is clearly a consequence of the strong repulsions between molecules, rather than of the weak attractions, which are what produced the peak in the dilute-gas $g(r)$.

In Table 7.1 there is tabulated the hard-sphere $g(r)$ at various number densities ρ as determined by computer simulation. The first column, labeled x, is the distance r in units of the hard-sphere diameter σ. The label V on the remaining columns is the volume per molecule, $1/\rho$, in units of the volume per molecule at close packing, $\sigma^3/\sqrt{2}$.

Exercise (7.1). The molar volume of liquid argon when it is in equilibrium with its vapor at 85 K is 28.39 cm³/mol. Which column in Table 7.1 most nearly corresponds to this density? (Take $\sigma = 3.4$ Å for argon.)

Solution. $\rho = N_0/v_m$ where N_0 is Avogadro's number and v_m is the given molar volume. Then $V = \sqrt{2}/(\rho\sigma^3) = \sqrt{2}v_m/(N_0\sigma^3) = \sqrt{2}(28.39)/[(6.02 \times 10^{23})(3.4 \times 10^{-8})^3] = 1.70$, so it is the column labeled $V = 1.7$ that corresponds to this density.

By definition, $\rho g(r)$ is the mean local density at the distance r from the center of any molecule. The volume of a shell of radius r and thickness dr is $4\pi r^2 dr$, so the average (infinitesimal) number of other molecules that have their centers at a distance between r and $r + dr$ from the center of any given molecule is $4\pi \rho r^2 g(r) dr$. Therefore the average total number N_R of other molecules whose centers are within the distance R of the center of any given molecule is

$$N_R = 4\pi\rho \int_0^R r^2 g(r) dr. \quad (7.1)$$

Table 7.1. *Hard-sphere g(r) at various number densities ρ*

x	$V = 3$	$V = 2$	$V = 1.7$	$V = 1.6$
1.00	2.07	3.30	4.36	4.95
1.04	1.92	2.77	3.44	3.73
1.08	1.78	2.36	2.68	2.89
1.12	1.65	2.03	2.17	2.23
1.16	1.53	1.76	1.79	1.76
1.20	1.43	1.55	1.49	1.44
1.24	1.35	1.37	1.27	1.20
1.28	1.28	1.23	1.10	1.01
1.32	1.21	1.10	0.97	0.89
1.36	1.16	1.01	0.87	0.80
1.40	1.10	0.94	0.80	0.73
1.44	1.06	0.88	0.74	0.69
1.48	1.02	0.84	0.70	0.66
1.52	0.99	0.81	0.69	0.65
1.56	0.97	0.78	0.69	0.65
1.60	0.95	0.79	0.69	0.67
1.64	0.94	0.79	0.71	0.69
1.68	0.92	0.80	0.75	0.74
1.72	0.92	0.82	0.80	0.80
1.76	0.92	0.85	0.86	0.87
1.80	0.92	0.88	0.91	0.96
1.84	0.92	0.91	0.98	1.03
1.88	0.93	0.96	1.04	1.10
1.92	0.95	1.00	1.09	1.17
1.96	0.97	1.06	1.15	1.22
2.00	0.99	1.10	1.22	1.28
2.04	1.00	1.14	1.27	1.31
2.08		1.15	1.26	1.28
2.12		1.14	1.22	1.23
2.16		1.13	1.17	1.16
2.20		1.10	1.12	1.10
2.24		1.08	1.06	1.03
2.28		1.06	1.01	0.99
2.32		1.04	0.98	0.95
2.36		1.02	0.96	0.91
2.40		1.00	0.94	0.88
2.44		0.98	0.92	0.87
2.48		0.97	0.91	0.87
2.52		0.96	0.92	0.87
2.56		0.96	0.92	0.89
2.60		0.95	0.92	
2.64		0.96		
2.68		0.96		
2.72		0.97		
2.76		0.98		
2.80		0.99		

Note: $x = r/\sigma$, $V = \sqrt{2}/(\rho\sigma^3)$.
Source: From B.J. Alder and C.E. Hecht, *J. Chem. Phys.* **50** (1969) 2032.

This number may then be found by measuring the area under a plot of $r^2 g(r)$ vs. r, between $r = 0$ and $r = R$, and then multiplying by $4\pi\rho$. When this is done for the hard-sphere fluid using the entries for $V = 1.7$ in Table 7.1, one finds for N_R, where R is the location of the first minimum in $g(r)$, a number close to 12, which is the number of spheres in the first shell of neighbors about any central sphere at close packing. About the same number is found from the experimental $g(r)$ as well, with the lesson that ordinary dense liquids are highly ordered locally – they are locally just slightly expanded from close packing – and that the typical molecule has many close neighbors – around 12 – within the range of the attractive forces. [It was remarked at the end of §6.2 that the attractive forces could be thought of as active out to about $r \simeq 1.5\sigma$, so to about 5.1 Å for argon; typically, then, out to the first minimum in $g(r)$.] That most molecules in a normal dense liquid have many close neighbors is what distinguishes the liquid from a dilute gas, and is a most important fact, which we shall wish to recall in the next section.

7.2 Equation of state of a liquid

We saw in §7.1 that the structure of a dense liquid is nearly the same as that of a fluid of hard, non-attracting spheres at a number density equal to that of the liquid – provided the diameter σ of the hard spheres is chosen equal to the distance of approach at which the strong repulsion between the real molecules sets in (Fig. 6.3). Each molecule of the liquid is surrounded by a large number of near neighbors to which it is attracted – but attracted more or less equally in all directions because the neighboring molecules are more or less uniformly spaced about the central one. The attractive *forces* therefore largely cancel each other; but the *potential energies* of attraction are additive, so the average net potential energy of attraction felt by each molecule is very large and negative – roughly -12ε, say, where ε is the depth of the well in the intermolecular interaction potential $\phi(r)$ (Fig. 6.1).

The picture that emerges, then, is that a dense liquid of simple, nearly spherical molecules is much like a fluid of hard spheres swimming in a deep but gradientless (i.e., spatially uniform) potential well – gradientless because the force is the negative of the gradient of the potential energy and we saw that the net attractive force would be practically nil. That there is little net attractive force on any molecule is consistent with the liquid's structure being practically the same as that of the reference hard-sphere fluid without attractions.

The depth of the potential well in which each molecule finds itself is proportional to the number of surrounding molecules contributing to it, which is in turn proportional to the number density ρ. Let the proportionality coefficient be called $-2a$; we shall later see that a is van der Waals's a [cf. Ex. (6.1) and the

7.2 Equation of state of a liquid

discussion following Eq. (6.24)]. It is independent of the temperature because in this picture the structure of the liquid around any given central molecule is the same as in the reference hard-sphere fluid, which is independent of the temperature (§7.1). The average potential energy of each molecule is now $-2a\rho$. To obtain the total potential energy in the liquid, which we shall again call E_{intermol} as in §6.3, we must multiply $-2a\rho$ by the total number of molecules N, and then divide by 2 so as not to have counted each interaction twice [cf. Eq. (6.10) and the discussion preceding it]:

$$E_{\text{intermol}} = -a\rho N \qquad (7.2)$$

Let $E_{\text{i.g.}}$ be again, as in §6.3, the energy of the corresponding ideal gas; i.e., what the total energy would have been had there been no intermolecular forces acting. For given N it is some function of T alone, equal to the sum of the translational and internal (intramolecular) energies. Then the total energy E of the liquid in this picture is

$$E = E_{\text{i.g.}} - a\rho N. \qquad (7.3)$$

The only temperature dependence is in $E_{\text{i.g.}}$ and the only volume dependence is in $\rho = N/V$. Therefore, for given fixed N,

$$\left(\frac{\partial E}{\partial V}\right)_T = a\left(\frac{N}{V}\right)^2 = a\rho^2, \qquad (7.4)$$

so from the thermodynamic identity (6.16),

$$\left[\frac{\partial(p/T)}{\partial(1/T)}\right]_\rho = -a\rho^2. \qquad (7.5)$$

Then

$$p/T = -a\rho^2/T + h(\rho) \qquad (7.6)$$

where $h(\rho)$ – not the same $h(\rho)$ as in (6.19)! – is again some function of ρ alone, independent of T. But when there are no attractive forces, only the hard-sphere repulsions, so that $a = 0$, the model is one of pure, non-attracting hard spheres, in which the pressure at the density ρ and temperature T is $p_{\text{h.s.}}$, say. That, then, identifies the function $h(\rho)$ in (7.6) as $p_{\text{h.s.}}/T$,

$$p_{\text{h.s.}} = Th(\rho), \qquad (7.7)$$

so that finally the equation of state of the liquid becomes

$$p = p_{\text{h.s.}} - a\rho^2. \qquad (7.8)$$

Note from (7.7) that the pressure of a fluid of non-attracting hard spheres at a given density is directly proportional to the absolute temperature. This is exact, and not dependent on any simplified picture or approximation. Because of the infinitely strong repulsions between hard spheres, in no configuration of such a fluid at any finite temperature do any of the spheres interpenetrate; in every realizable configuration every distance r between sphere centers is greater than σ and the total interaction energy is 0. The total energy of the fluid of non-attracting hard spheres is then just $E_{h.s.} = E_{i.g.}$, which for given N depends on T alone and is independent of V. Therefore $(\partial E_{h.s.}/\partial V)_T = 0$, or $[\partial(p_{h.s.}/T)/\partial(1/T)]_\rho = 0$, from (6.16). Therefore $p_{h.s.}/T$ must be some function $h(\rho)$ of ρ alone, independent of T, as asserted in (7.7).

Equations (7.7) and (7.8) together imply that, for ordinary dense liquids, isochores (curves of constant density) in the pressure–temperature plane should be straight lines. This key prediction of the theory is found to be very well borne out. In an authoritative assessment* we may read [here γ_V stands for $(\partial p/\partial T)_V$ or $(\partial p/\partial T)_\rho$],

This graph is always close to a straight line .... For example, Amagat's measurements ... for ethyl ether at 20 °C give the following figures for the slope and curvature of the isochores

$$\left(\frac{\partial p}{\partial T}\right)_V = 9.0 \text{ bar K}^{-1} \qquad \left(\frac{\partial^2 p}{\partial T^2}\right)_V = -0.009 \text{ bar K}^{-2}$$

Thus, in a typical experiment... in which the temperature is raised by 5 K ... (t)he change in γ_V ... would be only 0.05 bar K^{-1}, about 0.6 per cent, and so the curvature of a plot of p against T would be almost undetectable.

Note that the van der Waals equation of state is of the form of (7.8) with (7.7). Using the molecular rather than molar forms of the constants a and b, i.e., replacing n by N and R by k, the van der Waals equation, quoted in Ex. (6.1), is

$$(p + a\rho^2)(1/\rho - b) = kT \qquad (7.9)$$

or

$$p = \frac{kT}{1/\rho - b} - a\rho^2. \qquad (7.10)$$

This would be exactly (7.8) if the function of density $h(\rho)$ in (7.7) were

$$h(\rho) = \frac{k}{1/\rho - b} = k\frac{\rho}{1 - b\rho} \quad \text{(vdW)}. \qquad (7.11)$$

---

* J.S. Rowlinson and F.L. Swinton, *Liquids and Liquid Mixtures*, 3rd edn. (Butterworths, 1982), §2.3, p. 28.

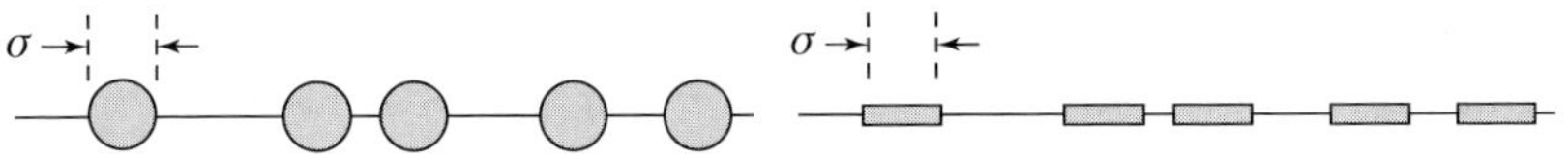

Fig. 7.3

How accurately does van der Waals's $h(\rho)$ in (7.11) represent the pressure–temperature ratio of a fluid of non-attracting rigid spheres? The answer is, "not very," for the hard-sphere fluid in three dimensions, although with $b = \sigma$ it is exact in one dimension (In one dimension we may equally well think of the particles as rigid rods of length σ; Fig. 7.3.) If the equation of state of the hard-sphere system were like that of an ideal gas in a volume diminished by the volumes v_0 of the molecules themselves, then we would expect $p_{\text{h.s.}}(V - Nv_0) = NkT$, or $p_{\text{h.s.}} = kT/(1/\rho - v_0)$. In one dimension the "volume" v_0 is the rod length (sphere diameter) σ, so this $p_{\text{h.s.}}$ is (7.7) with (7.11), with b identified as σ. As remarked, it is correct in one dimension. For example, it has the pressure $p_{\text{h.s.}}$ becoming infinite, as it should, as the spheres or rods become close packed on the line; i.e., as $N \to V/\sigma$, which is $1/\rho \to \sigma$. At low densities it implies

$$p_{\text{h.s.}}/kT = \rho/(1 - \sigma\rho) = \rho(1 + \sigma\rho + \sigma^2\rho^2 + \cdots) \quad \text{(1-dim.)}, \quad (7.12)$$

so that, by (6.2), the second virial coefficient is just $B = \sigma$, which is correct [Ex. (7.2)].

Exercise (7.2). Show that the second virial coefficient B of the one-dimensional fluid of hard rods of length σ is $B = \sigma$.

Solution. The formula (6.21) for the second virial coefficient in terms of the intermolecular potential $\phi(r)$ is for a three-dimensional fluid. The $2\pi r^2 dr$ in (6.21) started as $\frac{1}{2} d\tau$, with $d\tau$ an element of three-dimensional volume, as in (6.10); the integration over angles in going from (6.10) to (6.11) then produced the $\frac{1}{2}(4\pi r^2)dr = 2\pi r^2 dr$ in (6.11) that carried through to the formula for B in (6.21). If we had not taken advantage of the spherical symmetry of the integrand to do the angle integrations, (6.21) would have been

$$B = \frac{1}{2}\int \left(1 - e^{-\phi(r)/kT}\right) d\tau,$$

the integration being over all elements of volume $d\tau$ in the infinite space. In this form the formula for B is general for a system in any number of dimensions, provided only that the potential energy of intermolecular interaction, $\phi(r)$, depends only on the distance r between molecular centers and not on orientation. In one dimension, with coordinate x, the "volume" element $d\tau$ is dx, while r, the distance between centers, is $|x|$. The integration is over the whole one-dimensional space from $x = -\infty$ to $x = +\infty$. For hard rods, $\phi = 0$ for $|x| > \sigma$ (see figure),

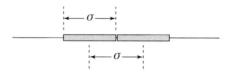

so the integrand vanishes for $x > \sigma$ and $x < -\sigma$; while $\phi = \infty$ for $|x| < \sigma$, so the integrand is 1 for x in the range $-\sigma < x < \sigma$. Then

$$B = \frac{1}{2}\int_{-\sigma}^{\sigma} dx = \sigma,$$

as required.

While the supposition $p_{\text{h.s.}} = kT/(1/\rho - v_0)$, which would include the van der Waals form (7.11) for $h(\rho)$ with b identified as the volume v_0 of a molecule, has thus proved correct in one dimension, it is very inaccurate in higher dimensions. The three-dimensional $p_{\text{h.s.}}$, or $h(\rho)$, has been determined by computer simulation using methods of the kind to be described in §§7.3 and 7.4, which were those also employed in determining the hard-sphere $g(r)$ tabulated in Table 7.1. A convenient analytical approximation to $h(\rho)$ in three dimensions is the now commonly used Carnahan–Starling formula,

$$h(\rho) = k\rho(1 + y + y^2 - y^3)/(1 - y)^3 \qquad (7.13)$$

where y is the number density ρ in units of the reciprocal of the sphere volume, $1/v_0 = [(4/3)\pi(\sigma/2)^3]^{-1} = 6/(\pi\sigma^3)$:

$$y = v_0\rho = \frac{\pi}{6}\sigma^3\rho. \qquad (7.14)$$

Equation (7.13) is of a form originally suggested by approximate analytical theories but modified to give correctly or nearly correctly the hard-sphere virial coefficients through the sixth. It reproduces the computer-simulation results quite well up to a density about 2/3 that of closest packing, i.e., for ρ up to about $(2/3)(\sqrt{2}/\sigma^3)$, at which point it continues to increase smoothly with increasing

7.2 Equation of state of a liquid

ρ while the simulations show an abrupt change, apparently connected with a transition from a disordered fluid state of the hard-sphere system to an ordered, crystalline state.

Exercise (7.3). According to Eqs. (7.7) and (7.11), van der Waals's approximation for the equation of state of a fluid of hard spheres is of the form

$$\frac{p_{\text{h.s.}}}{kT} = \frac{\rho}{1 - b\rho}$$

with b some measure of the volume of the sphere; while according to (7.13) and (7.14) a more accurate equation of state for hard spheres is

$$\frac{p_{\text{h.s.}}}{kT} = \rho \frac{1 + y + y^2 - y^3}{(1 - y)^3}$$

where $y = v_0 \rho$ with v_0 the sphere volume.

(a) Show that if we identify b with $4v_0$, the two expressions for $p_{\text{h.s.}}$ will agree at low density to terms of order ρ^2 but will differ in their terms of order ρ^3 and higher.

(b) With $b = 4v_0$, plot $v_0 p_{\text{h.s.}}/kT$ vs. $y (= v_0 \rho)$ for both equations. Note where each becomes infinite. What is the value of y that corresponds to the density of liquid argon at its triple point?

Solution.

(a) Expanding in powers of ρ, and noting the expansion $(1 - y)^{-3} = 1 + 3y + 6y^2 + \cdots$, we obtain from the two alternative forms of $p_{\text{h.s.}}$ above, respectively,

$$p_{\text{h.s.}}/kT = \rho(1 + b\rho + b^2\rho^2 + \cdots) = \rho(1 + 4v_0\rho + 16v_0^2\rho^2 + \cdots)$$

and

$$p_{\text{h.s.}}/kT = \rho(1 + v_0\rho + v_0^2\rho^2 - v_0^3\rho^3)(1 + 3v_0\rho + 6v_0^2\rho^2 + \cdots)$$
$$= \rho(1 + 4v_0\rho + 10v_0^2\rho^2 + \cdots).$$

We see that they agree in their terms of order ρ and ρ^2 but disagree in their terms of order ρ^3 and higher. The first expression for $p_{\text{h.s.}}/kT$ yields higher values than the second as soon as the density is great enough for the terms of order ρ^3 to become important.

(b) Let us abbreviate $v_0 p_{\text{h.s.}}/kT$ by Y, and call it Y_{vdW} and Y_{CS} (for Carnahan–Starling) for the first and second forms of $p_{\text{h.s.}}$ above, respectively.

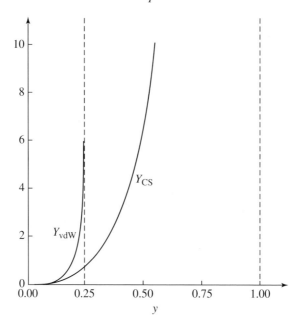

Then, with $b = 4v_0$ and $v_0 \rho = y$,

$$Y_{vdW} = y/(1 - 4y)$$
$$Y_{CS} = y(1 + y + y^2 - y^3)/(1 - y)^3.$$

These are plotted as functions of y in the accompanying figure. We see that Y_{vdW} lies above Y_{CS} for small y, as anticipated in (a). Y_{vdW} becomes infinite at $y = 1/4$ and is unphysical (negative) at higher y. Y_{CS} becomes infinite at $y = 1$. From Exercise (7.1) we know that for liquid argon near its triple point $\sqrt{2}/(\rho \sigma^3) = 1.70$. But $v_0 = (4\pi/3)(\sigma/2)^3$, so at that point $y = v_0 \rho = (\pi/6)\rho\sigma^3 = (\pi/6)(\sqrt{2}/1.70) = 0.436$. There, Y_{vdW} is unphysical while Y_{CS} still makes sense and is known to be quite accurate.

The equation of state (7.8) with a realistic three-dimensional $p_{h.s.} = Th(\rho)$ such as that in (7.13) gives a good qualitative account of ordinary dense liquids. We have noted that it is essentially the van der Waals equation of state but now with a three-dimensional rather than one-dimensional $p_{h.s.}$. It was crucial to the argument that led to (7.8) that the fluid be dense, as in the liquid near its triple point, so that each molecule has many neighbors and is attracted with nearly equal force in every direction. It would not be expected, then, to be a particularly accurate equation of state for an imperfect gas, or even for a liquid in the neighborhood of its critical point, where the density is only about one-third of what it is at the triple point. In the dilute gas, for example, (7.8) with (7.13) implies the same second virial coefficient $B = b - a/kT$ as does the van der Waals equation, with $b = 4v_0$ [Exercise (7.3); cf. Ex. (6.1)]; whereas we saw in Chapter 6, in the discussion following Ex. (6.2), that while this represents $B(T)$ reasonably well at high temperatures, it fails badly at low temperatures.

The van der Waals-like equation of state (7.8), and especially the physical picture that underlies it, is the beginning of the modern theory of liquids.* There are several directions in which it has been extended and improved, which we shall now examine.

If we continue to assume that the potential energy in the liquid is the sum of pair interactions $\phi(r)$ and continue to identify the radial distribution function $g(r)$ with that in the hard-sphere fluid of the same density, $g_{h.s.}(r)$, then (7.2) is merely an approximation to the energy of intermolecular interaction, $E_{intermol}$, which according to (6.10) should really then be

$$E_{intermol} = \frac{1}{2} N\rho \int g_{h.s.}(r)\phi(r) d\tau. \qquad (7.15)$$

What additional approximation or simplification was implicitly made in taking (7.2) for $E_{intermol}$ instead of the more general and more obvious (7.15)? Since $g_{h.s.}(r)$ is identically 0 for $r < \sigma$, the integral in (7.15) need be evaluated only for $r > \sigma$. If we make the simplifying assumption that the oscillations in $g_{h.s.}(r)$ about the value 1 for $r > \sigma$ are unimportant (cf. Fig. 7.2), and simply replace $g_{h.s.}(r)$ in (7.15) by 0 for $r < \sigma$ and by 1 for $r > \sigma$, we obtain

$$E_{intermol} \simeq \frac{1}{2} N\rho \int_{r>\sigma} \phi(r) d\tau. \qquad (7.16)$$

This is precisely (7.2), now with the identification of the van der Waals constant a in terms of the pair potential $\phi(r)$,

$$a = -\frac{1}{2} \int_{r>\sigma} \phi(r) d\tau. \qquad (7.17)$$

For $r > \sigma$, that is, outside the hard core of the potential, in this model of attracting hard spheres the potential $\phi(r)$ is the attractive part alone, so (7.17) may be interpreted as saying that a is minus one-half the integral over all space of the attractive component of the intermolecular potential. That attractive component of ϕ is negative, so the resulting a is positive. For the square-well

---

* D. Chandler, J.D. Weeks, and H.C. Andersen, "Van der Waals Picture of Liquids, Solids, and Phase Transformations," *Science* **220** (1983) 787.

potential in Fig. 6.5 we find from (7.17),

$$\begin{aligned} a &= -\frac{1}{2} \cdot 4\pi \int_{\sigma}^{\infty} r^2 \phi(r) dr \\ &= 2\pi\varepsilon \int_{\sigma_1}^{\sigma_2} r^2 dr \\ &= (2\pi/3)\varepsilon(\sigma_2^3 - \sigma_1^3), \end{aligned} \tag{7.18}$$

as we had already concluded in the previous chapter, Eq. (6.24).

An obvious way to improve the theory that was based on the assumption (7.2) for E_{intermol} would be to replace (7.2) by the more accurate (7.15), and this is now often done. The disadvantage of doing that is that one then loses the elegance and simplicity both of (7.2) and of the equation of state (7.8), and must rely on numerical computation to obtain $g_{\text{h.s.}}(r)$ over a range of densities and to evaluate the integral in (7.15) and related integrals for E_{intermol} and for the pressure. But with facilities for high-speed computing now ubiquitous that is hardly a problem.

Another direction in which the theory has been improved is to take σ to be a temperature-dependent hard-sphere diameter, to take account of the fact that the repulsive wall of the intermolecular potential is not truly infinitely steep; cf. Fig. 6.1. Another strategy is to base the theory not on a reference hard-sphere fluid but on a different hypothetical reference fluid, in which the molecules interact only via the repulsive component of $\phi(r)$, which is the part between $r = 0$ and $r = r_0$, where $d\phi/dr < 0$ (Fig. 6.1); and then, as before, to assume that the structure of the liquid is the same as that of the reference fluid. This, in effect, is what it means to assume a hard-sphere reference fluid with a temperature-dependent σ, provided the effective σ is related judiciously to the repulsive component of $\phi(r)$.

Finally, there are theories in which, once the intermolecular forces are given, the properties of the liquid are in principle calculated exactly, without approximation. These are by the methods of computer simulation, which are sketched in the next two sections.

7.3 Computer simulation: molecular dynamics

If we knew the intermolecular forces exactly – for example, if they were the sum of pair interactions each derived from the same known, spherically symmetric interaction potential $\phi(r)$ – and if we had a powerful enough computer, we could solve the equations of motion and thus know the locations and the velocities of all the molecules in our system at all times. We would thus have simulated, by

7.3 Computer simulation: molecular dynamics

computer, the detailed molecular dynamics of a real fluid. With that information about the positions and velocities of the molecules we could then calculate all the macroscopic properties of our system as appropriate averages over time or over the molecules. This is the technique of *molecular dynamics*, one of the two main methods of *computer simulation* and the subject of this section. The alternative method is the subject of §7.4.

If the model system contained too many molecules the computer's capacity to follow its time evolution would be exceeded; if it contained too few, it would no longer realistically simulate a macroscopic system. In every computer simulation the choice of the number of molecules is a compromise between those two conflicting requirements. As of the year 2000, simulations are almost routinely done with several hundred or a thousand particles, less routinely with several thousand. The data in Table 7.1 were obtained by molecular dynamics for a system of only 108 hard spheres in runs of 100,000 collisions at each density.

Even with 1000 particles, and all the more so with only one or two hundred, "finite-size" effects may be severe. In a real macroscopic system, with short-range forces, the fact that the condition of the fluid at the container walls is not representative of the interior is of no consequence, for at any moment only a negligible fraction of the sample is in that unrepresentative condition. That is because the effect of the surface extends only over some *micro*scopic distance ℓ. Then if L is the *macro*scopic linear dimension of the container, so that the surface area is of order L^2, the volume of material affected by the surface is only of order ℓL^2, which is less, by the factor $\ell/L \approx 10^{-8}$, than the total volume L^3.

But now consider a simulation with 1000 hard spheres of diameter σ in a cubical box of edge length 11.5σ. Since the center of a sphere cannot come closer to a wall than $\frac{1}{2}\sigma$ the sphere *centers* are in effect confined to a box of edge length 10.5σ. The number density ρ of sphere centers is then $1000/(10.5\sigma)^3 = 0.86/\sigma^3$. [This is just a little greater than the number density at "$V = 1.7$" in Table 7.1, which is $\sqrt{2}/(1.7\sigma^3) = 0.83/\sigma^3$, and is about 60% of the density of closest packing, which is $\rho = \sqrt{2}/\sigma^3 = 1.41/\sigma^3$.] Then the number of these spheres whose centers are within, say, one-and-a-half sphere diameters of the wall, is around 470 – almost half the sample [Ex. (7.4)].

Exercise (7.4). In a simulated liquid consisting of 1000 spheres of diameter σ in a cubical box of edge length 11.5σ, how many of the spheres have their centers within $1\frac{1}{2}$ sphere diameters of the wall?

Solution. In order not to have its center within $1\frac{1}{2}$ sphere diameters of the wall a sphere must have its center in a cube of edge length 8.5σ inside the box. Since the number density of the sphere centers is $1000/(10.5\sigma)^3$ (see text), the number inside the smaller cube, of edge length 8.5σ, will be close to $[1000/(10.5\sigma)^3](8.5\sigma)^3 \simeq 530$. The rest of the 1000 spheres, so around 470, will have their centers within $1\frac{1}{2}$ sphere diameters of the wall.

In practice, such finite-size effects are minimized – although certainly not eliminated – by imposing "periodic boundary conditions" during the calculation. Figure 7.4 shows, schematically in two dimensions, such a simulation volume. It is shown surrounded on all sides by an infinite periodic array of exact replicas of itself, with the molecules (white) in exactly the same positions in the replica boxes as are the molecules (black) in the central simulation box itself. The molecules do not interact with the walls but those in the central simulation box are taken to interact with the image molecules in the replica boxes as well as with each other. Because of the exact replication, as a molecule – such as the one in Fig. 7.4 marked with an arrow, representing its velocity – leaves the central box it is immediately replaced by a replica of itself entering through the corresponding point of the opposite side. The walls have thus been made transparent to the particles, and so affect them little, yet the number of particles in the box remains fixed. It might seem as though we now have an infinite system, but the periodicity imposed by the replication is of finite wavelength equal to the box size, so there are still finite-size effects – although minimized because of the transparency of the walls.

After such a molecular dynamics simulation has proceeded for a long enough time the condition of the system – the locations of the molecules and the

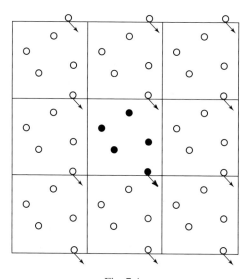

Fig. 7.4

velocity distribution – is that of an equilibrium system. The energy E is constant throughout the simulation because it is that of an isolated mechanical system; it remains at whatever value was set by the arbitrarily chosen initial configuration and velocities. Indeed, constancy of the energy is an important criterion by which one judges the reliability of the simulation.

The thermodynamic functions of the system may be calculated once the equilibrium configuration and velocities have been achieved. For example, the temperature $T(E, N, V)$ for the fixed total energy E, number of particles N, and box volume V may be inferred from the mean translational kinetic energy $\bar{K}$ of the molecules in the equilibrium system: $\bar{K} = \frac{3}{2}kT$. [See below, Eq. (7.25) and the discussion that accompanies it.] This could be equally well an average over all the molecules at any one time or a long-time average for any one molecule. This equivalence would be another test of the reliability of the simulation. Note that at liquid densities $N\bar{K}$ is very different from E: a large part of E is the potential energy E_{intermol}.

Since a typical equilibrium configuration is achieved in the simulation, the radial distribution function $g(r)$ may be inferred from the positions of the molecules in that configuration. If we consider a spherical shell of inner radius R and outer radius $R + \Delta R$ centered at any one molecule (Fig. 7.5; the dots are molecule centers), then it follows from Eq. (7.1) that the number ΔN_R of molecules whose centers lie in that shell is on average

$$\Delta N_R = N_{R+\Delta R} - N_R$$
$$= 4\pi \rho \int_R^{R+\Delta R} r^2 g(r) dr$$
$$\simeq 4\pi \rho R^2 g(R) \Delta R \qquad (7.19)$$

where the approximation (7.19) would be accurate as long as ΔR was small enough. If R in (7.19) were replaced, say, by $R + \frac{1}{2}\Delta R$, it would be even

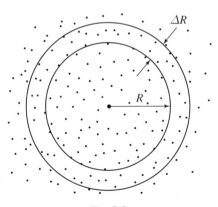

Fig. 7.5

more accurate. From the configuration achieved in the simulation the number ΔN_R of molecules in the shell may be counted, and the overall number density $\rho = N/V$ is known, so the value of the radial distribution function $g(r)$ at $r = R$ (or $R + \frac{1}{2}\Delta R$) may then be calculated from (7.19); and this may be done for many closely spaced values of r.

For good statistics the process should be repeated with each of the N molecules in turn taken to be the central molecule in any one equilibrium configuration, and it should be done also for many different times; i.e., many different configurations, each typical of equilibrium. Thus averaged over many molecules and many configurations, the resulting $g(r)$ is a good approximation to the $g(r)$ of the *model* system. How good an approximation that, in turn, is to the $g(r)$ of a real fluid is a different question.

Another important quantity one would wish to derive from a molecular dynamics simulation is the pressure p of the fluid at the given density ρ and energy E. Since the temperature T at that ρ and E is derivable from $\bar{K}$, as described above, calculating p for various ρ and E would then in effect give us $p(\rho, T)$, the form in which one would usually wish to know the equation of state.

If the simulation were done in a box without periodic boundary conditions, and if one took account of the collisions of the molecules with the walls of the box, one could calculate p just as in the kinetic theory of gases; i.e., as the total momentum change suffered by the molecules that reflect off the walls, per unit area per unit time. There is another way to obtain p in the simulation, however, that is more convenient and numerically more accurate. It requires only the typical equilibrium configurations – the same information as in $g(r)$ – and knowledge of the interaction potential $\phi(r)$. This method is via the virial theorem of mechanics, which we referred to in passing in §6.1.

Think for the moment of the molecules as being in a container with real walls, not the periodic box of Fig. 7.4. Let the vector $\mathbf{r}(t)$, measured as a displacement from an arbitrary fixed origin, be the location of the center of a given molecule at the time t. The molecule's velocity is $d\mathbf{r}/dt$ and its kinetic energy is

$$K = \frac{1}{2}m \frac{d\mathbf{r}}{dt} \cdot \frac{d\mathbf{r}}{dt} \tag{7.20}$$

where m is the mass of the molecule and the dot product is the scalar product of $d\mathbf{r}/dt$ with itself, giving the square of its magnitude. This expression for K may be rearranged to

$$K = \frac{1}{2}m \frac{d}{dt}\left(\mathbf{r} \cdot \frac{d\mathbf{r}}{dt}\right) - \frac{1}{2}m\mathbf{r} \cdot \frac{d^2\mathbf{r}}{dt^2}, \tag{7.21}$$

as may be verified by carrying out the indicated differentiation in the first

7.3 Computer simulation: molecular dynamics

term. The second derivative $d^2\mathbf{r}/dt^2$ is the particle's acceleration, and so when multiplied by the particle's mass m it is the net force $\mathbf{F}$ on the particle:

$$K = \frac{1}{2}m\frac{d}{dt}\left(\mathbf{r} \cdot \frac{d\mathbf{r}}{dt}\right) - \frac{1}{2}\mathbf{r} \cdot \mathbf{F}. \tag{7.22}$$

The quantity $\mathbf{r} \cdot \mathbf{F}$ – the scalar product of the force on the particle with the particle's position vector – is called the *virial* of the force on the particle.

The product $\mathbf{r} \cdot d\mathbf{r}/dt$ remains bounded (finite) for all times: $|\mathbf{r}|$ is bounded because the particle is confined to a box, and $|d\mathbf{r}/dt|$ is bounded because the system's total energy E is a fixed constant, which would not be possible if any particle's velocity increased without bound. But the long-time average $\overline{df/dt}$ of the derivative of any bounded function $f(t)$ of the time is necessarily 0:

$$\begin{aligned}\overline{df/dt} &= \lim_{\tau \to \infty} \frac{1}{\tau}\int_0^\tau (df/dt)dt \\ &= \lim_{\tau \to \infty} \frac{f(\tau) - f(0)}{\tau} = 0,\end{aligned} \tag{7.23}$$

the limit as $\tau \to \infty$ being 0 because $f(\tau)$, by supposition, remains bounded as $\tau \to \infty$. This is really just the common-sense remark that if the long-time average of the derivative of $f(t)$ were *not* 0 then $f(t)$ would be increasing or decreasing at some non-0 average rate for all time and would therefore increase or decrease without bound as time went on.

Since we remarked above that $\mathbf{r} \cdot d\mathbf{r}/dt$ remains bounded for all time, and since we now know that this requires the long-time average of its time derivative to vanish, we obtain from (7.22), upon taking the long-time average of both sides,

$$\bar{K} = -\frac{1}{2}\overline{\mathbf{r} \cdot \mathbf{F}}; \tag{7.24}$$

i.e., the long-time average kinetic energy of any of the particles is $-\frac{1}{2}$ of the long-time average of the virial of the net force on that particle. This is the virial theorem.

In connection with the equipartition law, §2.5, we saw that any "square term" – whether it be the square of a velocity component or the square of a coordinate – that occurs as its own separate term in the Hamiltonian, not coupled to any of the other degrees of freedom, contributes $\frac{1}{2}kT$ to the average energy. This assumed only that the temperature was high enough for the degree of freedom that contributes that term to be treated classically, and that would certainly be true of the translational degrees of freedom of the molecules of practically any fluid at practically any temperature of interest. Thus, with each molecule having three translational degrees of freedom,

$$\bar{K} = \frac{3}{2}kT. \tag{7.25}$$

This is true not only for an ideal gas, as in Chapter 2, but for any system, including imperfect gases and liquids, for which the translational motion of the molecules is essentially classical (non-quantum mechanical). Indeed, not only the mean kinetic energy but the whole velocity distribution function is the same in an imperfect gas or in a liquid as it is in an ideal gas, as we already remarked in Chapter 1 (§1.2). That is because the molecules' translational velocities occur in strictly separable square terms in the Hamiltonian, not coupled to the coordinates or to each other. Thus, (7.24) is now

$$\overline{\mathbf{r} \cdot \mathbf{F}} = -3kT. \tag{7.26}$$

The net force $\mathbf{F}$ on any particle is made up partly of the forces $\mathbf{F}_{other}$ that all the other molecules in the system exert on it, and partly of the force $\mathbf{F}_{wall}$ that the wall exerts on it to keep it from leaving the container. Thus, from (7.26),

$$\overline{\mathbf{r} \cdot \mathbf{F}_{other}} + \overline{\mathbf{r} \cdot \mathbf{F}_{wall}} = -3kT. \tag{7.27}$$

It would be a good guess that the $\overline{\mathbf{r} \cdot \mathbf{F}_{wall}}$ term in (7.27), having nothing to do with the intermolecular forces, would be the same as in the corresponding ideal gas, for which $kT = pV/N$; i.e., that in general, even when there are intermolecular forces acting,

$$\overline{\mathbf{r} \cdot \mathbf{F}_{wall}} = -3pV/N. \tag{7.28}$$

That this really is so may be seen as follows. Any molecule near the wall is just as likely to be near any one element of area $d\sigma$ of the wall as near any other of equal size; and the pressure p, which is the force per unit area that the molecules of the fluid exert on the wall, is uniform over the wall. Further, the time average for any one molecule is the same as the average over all the molecules at any one instant. Therefore, as we shall see momentarily, the *total* of $\mathbf{r} \cdot \mathbf{F}_{wall}$, summed over all N molecules of the fluid, is

$$N \overline{\mathbf{r} \cdot \mathbf{F}_{wall}} = -p \int_S \mathbf{r} \cdot \mathbf{n} d\sigma \tag{7.29}$$

where the integral is over the whole surface S of the container and where the vector $\mathbf{n}$ is the unit outward normal to the surface at the point $\mathbf{r}$ at which the

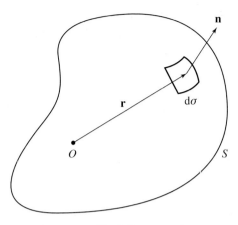

Fig. 7.6

element of area dσ is located. (Fig. 7.6; O is the arbitrary origin from which **r** is measured.) To see where (7.29) comes from note that p**n**dσ is the total force exerted on the element of area dσ of the wall by all the molecules in the container (pdσ being its magnitude and **n** its direction), and so is the negative of the force that element of the wall exerts on the molecules; and the molecules that feel that force are those that are at the position **r** at which dσ is located.

From Green's theorem of the vector calculus, the integral of the normal component **n** · **u** of any vector **u**, integrated over a closed surface S, is the same as the integral of the divergence of **u**, i.e., the integral of $\nabla \cdot \mathbf{u} = \partial u_x/\partial x + \partial u_y/\partial y + \partial u_z/\partial z$, integrated through the volume V enclosed by S. Thus, with dτ an element of volume at the position **r** in V,

$$\int_S \mathbf{r} \cdot \mathbf{n} d\sigma = \int_V \nabla \cdot \mathbf{r} d\tau. \qquad (7.30)$$

But the components of **r** are x, y, and z, so $\nabla \cdot \mathbf{r} = 3$, the dimensionality of space. Therefore the right-hand side of (7.30) is just 3 times the volume V of the container, and so the left side is, too:

$$\int_S \mathbf{r} \cdot \mathbf{n} d\sigma = 3V. \qquad (7.31)$$

From (7.29), then,

$$N \overline{\mathbf{r} \cdot \mathbf{F}_{\text{wall}}} = -3pV, \qquad (7.32)$$

exactly as guessed in (7.28).

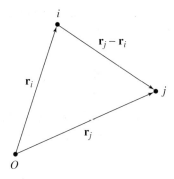

Fig. 7.7

Putting (7.32) into (7.27), we obtain a formula by which the pressure p may be evaluated in a molecular dynamics simulation:

$$\begin{aligned} p &= -\frac{1}{3}\frac{N}{V}\overline{\mathbf{r}\cdot\mathbf{F}_{\text{wall}}} \\ &= -\frac{1}{3}\rho(-3kT - \overline{\mathbf{r}\cdot\mathbf{F}_{\text{other}}}) \\ &= \rho\left(kT + \frac{1}{3}\overline{\mathbf{r}\cdot\mathbf{F}_{\text{other}}}\right), \end{aligned} \qquad (7.33)$$

where, as usual, ρ is the number density N/V. Note that if the molecules do not interact with each other then there is no $\mathbf{F}_{\text{other}}$ and (7.33) becomes $p = \rho kT$, the ideal-gas law.

Equation (7.33) may be made more explicit if we assume that the total potential energy of interaction is the sum of pair interactions, $\phi(r)$. Figure 7.7 shows the centers of a pair of molecules i and j at the positions $\mathbf{r}_i$ and $\mathbf{r}_j$ relative to an arbitrary fixed origin O. The distance r_{ij} between the molecules is $r_{ij} = |\mathbf{r}_j - \mathbf{r}_i|$. Let $\mathbf{F}_{j\,\text{on}\,i}$ be the force that molecule j exerts on molecule i. If $\phi'(r_{ij}) > 0$ [the prime means differentiation with respect to the indicated argument: $\phi'(r) = d\phi(r)/dr$] then the force of j on i or of i on j is one of attraction; if $\phi'(r_{ij}) < 0$ it is one of repulsion (cf. Fig. 6.1). The magnitude of the force $\mathbf{F}_{j\,\text{on}\,i}$ is $|\phi'(r_{ij})|$, while its direction is that of the unit vector $(\mathbf{r}_j - \mathbf{r}_i)/r_{ij}$ if it is attractive or $(\mathbf{r}_i - \mathbf{r}_j)/r_{ij}$ if it is repulsive. Thus, in either case,

$$\mathbf{F}_{j\,\text{on}\,i} = \phi'(r_{ij})\frac{\mathbf{r}_j - \mathbf{r}_i}{r_{ij}} \qquad (7.34)$$

and therefore

$$\mathbf{r}_i \cdot \mathbf{F}_{j\,\text{on}\,i} = \phi'(r_{ij})\mathbf{r}_i \cdot \frac{\mathbf{r}_j - \mathbf{r}_i}{r_{ij}}. \qquad (7.35)$$

We obtain the virial of the total force that all the other molecules of the fluid exert on the molecule i by summing (7.35) over all $j = 1, \ldots, N$ other than $j = i$:

$$\mathbf{r}_i \cdot \mathbf{F}_{\text{others on }i} = \sum_{j(\neq i)} \phi'(r_{ij}) \mathbf{r}_i \cdot \frac{\mathbf{r}_j - \mathbf{r}_i}{r_{ij}}. \quad (7.36)$$

The $\overline{\mathbf{r} \cdot \mathbf{F}_{\text{other}}}$ in (7.33) is just the quantity in (7.36) averaged over all the molecules i of the fluid. We obtain that, then, by summing over i and dividing by N:

$$\overline{\mathbf{r} \cdot \mathbf{F}_{\text{other}}} = \frac{1}{N} \sum_i \sum_{\substack{j \\ (i \neq j)}} \phi'(r_{ij}) \mathbf{r}_i \cdot \frac{\mathbf{r}_j - \mathbf{r}_i}{r_{ij}}. \quad (7.37)$$

But now on the right-hand side i and j are both dummy summation indices and we may at will interchange their roles. Thus, (7.37) is equally well

$$\overline{\mathbf{r} \cdot \mathbf{F}_{\text{other}}} = \frac{1}{N} \sum_i \sum_{\substack{j \\ (i \neq j)}} \phi'(r_{ij}) \mathbf{r}_j \cdot \frac{\mathbf{r}_i - \mathbf{r}_j}{r_{ij}}. \quad (7.38)$$

(The *distance* r_{ij} is symmetric: $r_{ij} = r_{ji}$.) Since $\overline{\mathbf{r} \cdot \mathbf{F}_{\text{other}}}$ is given simultaneously by the right-hand sides of (7.37) and (7.38) it is also half their sum, so

$$\overline{\mathbf{r} \cdot \mathbf{F}_{\text{other}}} = \frac{1}{2N} \sum_i \sum_{\substack{j \\ (i \neq j)}} \phi'(r_{ij}) \frac{1}{r_{ij}} (2\mathbf{r}_i \cdot \mathbf{r}_j - \mathbf{r}_i \cdot \mathbf{r}_i - \mathbf{r}_j \cdot \mathbf{r}_j). \quad (7.39)$$

But

$$\begin{aligned} r_{ij}^2 &= |\mathbf{r}_j - \mathbf{r}_i|^2 \\ &= (\mathbf{r}_j - \mathbf{r}_i) \cdot (\mathbf{r}_j - \mathbf{r}_i) \\ &= \mathbf{r}_j \cdot \mathbf{r}_j + \mathbf{r}_i \cdot \mathbf{r}_i - 2\mathbf{r}_i \cdot \mathbf{r}_j, \end{aligned} \quad (7.40)$$

so (7.39) is

$$\overline{\mathbf{r} \cdot \mathbf{F}_{\text{other}}} = -\frac{1}{2N} \sum_i \sum_{\substack{j \\ (i \neq j)}} r_{ij} \phi'(r_{ij}). \quad (7.41)$$

In words, to evaluate the quantity $\overline{\mathbf{r} \cdot \mathbf{F}_{\text{other}}}$ required in (7.33), sum $r_{ij}\phi'(r_{ij})$ over all pairs of molecules i and j and then divide by $-N$. (The double sum counts every pair twice.)

The molecular dynamics simulation gives typical equilibrium configurations from any of which one may read off all the distances r_{ij} between all pairs of molecules i and j, and thus, with the function $\phi(r)$ given, evaluate (7.41). There are $\frac{1}{2}N(N-1)$ pairs i, j, but in practice one need count only the much smaller number of them, of order N, for which the distance r_{ij} is small enough for $r_{ij}\phi'(r_{ij})$ not to be negligible, since $\phi'(r)$ ultimately falls off rapidly with increasing r. Once $\overline{\mathbf{r}\cdot\mathbf{F}_{\text{other}}}$ has thus been evaluated, the pressure follows from (7.33). Repeating the simulation for a range of densities ρ and energies E then gives the equation of state $p = p(\rho, T)$, as remarked earlier.

For analytical, as distinct from numerical, convenience, one often re-expresses (7.41) as an integral over a continuous distribution of molecules rather than as a sum over molecules at discrete positions. Any one of the N molecules i in the summation in (7.41) could be considered a central one from which to measure distance r, and the (infinitesimal) average number of other molecules j in a shell of inner radius r and outer radius $r + dr$ centered at i is $4\pi\rho r^2 g(r)dr$, from (7.19); so for a large enough system, (7.41) is equivalent to

$$\overline{\mathbf{r}\cdot\mathbf{F}_{\text{other}}} = -2\pi\rho \int_0^\infty r^3 g(r)\, \phi'(r)\, dr. \tag{7.42}$$

Then from (7.33),

$$p = \rho\left[kT - \frac{2}{3}\pi\rho \int_0^\infty r^3 g(r)\, \phi'(r)\, dr\right]. \tag{7.43}$$

This is often called the virial theorem of statistical mechanics.

When the potential $\phi(r)$ has an infinitely hard core, such as with hard spheres or with the square-well potential of Fig. 6.5, its derivative $\phi'(r)$ is not well defined and one requires variants of (7.41) or (7.43) obtained from (7.43) by a mathematical limiting process. As seen earlier, in Eq. (6.6), at low densities the product of $g(r)$ and $\exp[\phi(r)/kT]$ is just 1 at all r. As the repulsive part of $\phi(r)$ becomes steeper and steeper, ultimately becoming a hard core, so that $\exp[\phi(r)/kT]$ becomes ∞ for $r < \sigma$, the value of $g(r)$ becomes compensatingly smaller and smaller, ultimately becoming identically 0 for $r < \sigma$, but in such a way that the product $y(r) = g(r)\exp[\phi(r)/kT]$ remains fixed at 1 for all r, whether r be less than, equal to, or greater than σ. When the fluid is not a dilute gas the function $y(r)$, which we continue to define as

$$y(r) = g(r)e^{\phi(r)/kT}, \tag{7.44}$$

no longer has the value 1 at all r, but it nevertheless remains a continuous function even when the repulsive core of $\phi(r)$ becomes infinitely hard; $g(r)$ just

7.3 Computer simulation: molecular dynamics

vanishes rapidly enough to compensate for the diverging $\exp[\phi(r)/kT]$. Their product, $y(r)$, varies perfectly smoothly as r passes through the value σ and has a well-defined limiting value, $y(\sigma)$, at that point, independently of whether r approaches σ from above or below.

Noting this, and observing also that $y(r) \, d \exp[-\phi(r)/kT] = -(1/kT)\phi'(r) \times y(r) \exp[-\phi(r)/kT] dr = -(1/kT)\phi'(r) g(r) \, dr$, we may rewrite (7.43) as

$$p = \rho k T \left[1 + \frac{2}{3} \pi \rho \int_{r=0}^{r=\infty} r^3 y(r) \, d\, e^{-\phi(r)/kT} \right]. \qquad (7.45)$$

Suppose now that this is the hard-sphere model itself, with $\phi(r) = \infty$ for $r < \sigma$ and $\phi(r) = 0$ for $r > \sigma$. Then

$$e^{-\phi(r)/kT} = \begin{cases} 0, & r < \sigma \\ 1, & r > \sigma. \end{cases} \qquad (7.46)$$

This is the unit step function shown in Fig. 7.8, with the step at $r = \sigma$. Its differential, which appears in the integral in (7.45), is then $\delta(r - \sigma) dr$, with $\delta(r - \sigma)$ the delta-function centered at $r = \sigma$. We can see that because, from the definition of a δ-function (familiar, for example, from quantum mechanics),

$$\int_0^r \delta(r' - \sigma) \, dr' = \begin{cases} 0, & r < \sigma \\ 1, & r > \sigma, \end{cases} \qquad (7.47)$$

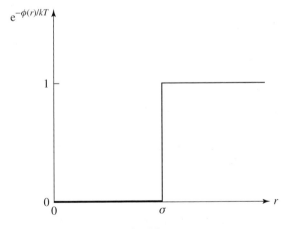

Fig. 7.8

which is the same as $\exp[-\phi(r)/kT]$ in (7.46), so $d\exp[-\phi(r)/kT]/dr = \delta(r-\sigma)$, as asserted. Equation (7.45) is then

$$p = \rho kT\left[1 + \frac{2}{3}\pi\rho\int_0^\infty r^3 y(r)\,\delta(r-\sigma)\,dr\right]$$

$$= \rho kT\left[1 + \frac{2}{3}\pi\rho\sigma^3 y(\sigma)\right] \qquad (7.48)$$

for hard spheres.

By the continutity of $y(r)$, we may evaluate $y(\sigma)$ by letting $r \to \sigma$ from either direction. Let $r \to \sigma$ from the $r > \sigma$ side. There $\exp[\phi(r)/kT]$ is 1; so from the definition of $y(r)$ in (7.44),

$$y(\sigma) = g(\sigma+), \qquad (7.49)$$

where $g(\sigma+)$ means the limiting value of $g(r)$ as $r \to \sigma$ from the $r > \sigma$ side. Note that while $y(r)$ is continuous at the surface of the hard core, $g(r)$ is not; for $r < \sigma$ we have $g(r) \equiv 0$, so what we might call $g(\sigma-)$ is 0, different from $g(\sigma+)$. This discontinuity in $g(r)$ at $r = \sigma$ is clearly visible in Fig. 7.2, where $g(\sigma-) = 0$ while $g(\sigma+) = 4.36$.

With $y(\sigma)$ now identified in (7.49), we obtain from (7.48)

$$p = \rho kT\left[1 + \frac{2}{3}\pi\rho\sigma^3 g(\sigma+)\right]$$

$$= \rho kT[1 + 4v_0\rho g(\sigma+)] \quad \text{(hard spheres)} \qquad (7.50)$$

where, as earlier (§7.2), $v_0 = \pi\sigma^3/6$ is the volume of a sphere of diameter σ.

We know that the structure of the hard-sphere fluid, and so its $g(r)$, depends only on the density but not on the temperature. Then (7.50) has the pressure equal to the temperature times a function $h(\rho)$ of the density alone, in agreement with (7.7), but now with an explicit formula for $h(\rho)$ in terms of the contact value $g(\sigma+)$ of the hard-sphere $g(r)$. This $g(r)$ is obtained in the simulations from (7.19), and its limiting value as $r \to \sigma+$ may be estimated by extrapolation from its values at slightly larger r. That, then, from (7.50), is the source of our knowledge of the hard-sphere $p_{\text{h.s.}}$ for use in the equation of state (7.8).

Exercise (7.5). At the density corresponding to "$V = 1.7$" in Table 7.1, by what factor does the pressure of the hard-sphere fluid exceed that of an ideal gas of the same density and temperature?

Solution. In the table, "$x = 1.00$" means $r = \sigma$; the entries in that line of the table are in fact the extrapolated $g(\sigma+)$. Then at $V = 1.7$, where

$\rho = \sqrt{2}/(1.7\sigma^3)$ and $v_0 \rho = (\pi\sigma^3/6)\sqrt{2}/(1.7\sigma^3) = 0.436$, we read $g(\sigma+) = 4.36$. Then from (7.50),

$$p_{\text{h.s.}} = \rho kT(1 + 4 \times 0.436 \times 4.36) = 8.6\rho kT.$$

At this density, $p_{\text{h.s.}}$ is 8.6 times the pressure of the corresponding ideal gas.

7.4 Computer simulation: Monte Carlo

This is a method of computer simulation in which the values of certain elements are determined by chance; it takes its name from that of the famed gambling resort. In its most common form it simulates a system of fixed temperature, rather than one of fixed energy as in the most usual form of molecular dynamics, §7.3. Like molecular dynamics, it generates equilibrium configurations. It is subject to the same finite-size effects as is molecular dynamics, and also to the same limitations on system size imposed by the limited capacity and speed of computers.

We shall for simplicity continue to assume that the potential energy of interaction of the molecules depends only on their locations, $\mathbf{r}_1, \ldots, \mathbf{r}_N$, and not on their orientations; thus, that the molecules are spherical or nearly so. That, at least implicitly, has been our assumption throughout. There is no major conceptual difficulty in treating non-spherical molecules, only more degrees of freedom to keep track of and greater awkwardness of notation. Let $W(\mathbf{r}_1, \ldots, \mathbf{r}_N)$ be the potential energy in the configuration $\mathbf{r}_1, \ldots, \mathbf{r}_N$. The further simplification of assuming that W is the sum of pair interactions $\phi(r_{ij})$,

$$W = \frac{1}{2} \sum_{\substack{i \ j \\ (i \neq j)}} \phi(r_{ij}), \tag{7.51}$$

required for formulas such as (7.15), (7.41), and (7.43), is also often made, although we need not assume it for now.

By the Boltzmann distribution law, the probability $P(\mathbf{r}_1, \ldots, \mathbf{r}_N) d\tau_1 \cdots d\tau_N$ of finding molecule 1 in the element of volume $d\tau_1$ at $\mathbf{r}_1$, at the same time finding molecule 2 in $d\tau_2$ at $\mathbf{r}_2$, etc., at the temperature T, is proportional to $\exp[-W(\mathbf{r}_1, \ldots, \mathbf{r}_N)/kT] d\tau_1 \cdots d\tau_N$. If these N molecules are in a volume V then the probability density P, properly normalized, is

$$P(\mathbf{r}_1, \ldots, \mathbf{r}_N) = \frac{e^{-W(\mathbf{r}_1, \ldots, \mathbf{r}_N)/kT}}{\int_V \cdots \int_V e^{-W(\mathbf{r}_1, \ldots, \mathbf{r}_N)/kT} d\tau_1 \cdots d\tau_N}, \tag{7.52}$$

where the denominator is the integral of the Boltzmann factor over all the possible positions $\mathbf{r}_i$ of all the molecules inside the volume V. Each integral over V is really a triple integral over the coordinates x_i, y_i, z_i, which are the components of the position vector $\mathbf{r}_i$ of molecule i. The element of volume $d\tau_i$ is $dx_i dy_i dz_i$.

The denominator in (7.52) is called the configuration integral. If divided by $N!$, so as not to count as different those configurations that differ from each other only by the interchange of identical particles (as in our treatment of the ideal gas, Chapter 2, §2.1), it would be the configurational factor in the total partition function Z of the system. Since the potential energy W depends only on the coordinates while the kinetic energy depends only on the velocities, they contribute independently and additively to the Hamiltonian, and so (in classical statistical mechanics, where one need not be concerned about the non-commutativity of coordinates and their conjugate momenta), they contribute multiplicatively to the partition function Z. The remaining factors in Z, then, besides the configuration integral (divided by $N!$), are those contributed by the translational velocities or momenta of the molecules, and any additional factors z_{int} from the molecules' internal degrees of freedom – rotational, vibrational, and electronic – assuming these to be independent of the configuration $\mathbf{r}_1, \ldots, \mathbf{r}_N$. Since they have nothing to do with the positions $\mathbf{r}_i$ of the molecules or the interaction energy W, these non-configurational contributions to Z are the same as in the corresponding ideal gas.

The velocities or momenta associated with the translational degrees of freedom of each molecule contribute the factor $(2\pi mkT/h^2)^{3/2}$ to Z, as seen in Chapter 2, Eq. (2.17). The remaining factor V in z_{trans} is configurational, contributed by the molecule's coordinates. That factor does depend, crucially, on there being no intermolecular interactions – no position-dependent W – and only for the ideal gas is it simply V. In the ideal gas the configurational factor in Z contributed by the N molecules together is V^N. It is exactly this V^N that is replaced by the configuration integral [the denominator of (7.52)] when there is a configuration-dependent W. If W were identically 0, as in an ideal gas, the multiple integral would be just the product of N independent integrals over V, and each would be just the volume V itself, so the configuration integral would be the same as the ideal-gas V^N. It is precisely because the configuration integral cannot, in general, be so decomposed, that a theory of the liquid state was so elusive for so long.

If $F(\mathbf{r}_1, \ldots, \mathbf{r}_N)$ is any function of the configuration, the mean value of F in the equilibrium fluid at temperature T, from (7.52), is

$$\bar{F} = \frac{\int_V \cdots \int_V F e^{-W/kT} d\tau_1 \cdots d\tau_N}{\int_V \cdots \int_V e^{-W/kT} d\tau_1 \cdots d\tau_N}. \tag{7.53}$$

For example, if F were W itself then $\bar{F}$ would be the mean intermolecular

7.4 Computer simulation: Monte Carlo

interaction energy, which we have been calling E_{intermol}. Adding the ideal-gas $E_{\text{i.g.}}$ to it, as in §6.3 or §7.2, would give us the total energy E. In principle, then, if we could evaluate the integrals in (7.53) we would know $E(N, V, T)$; and similarly $\bar{F}(N, V, T)$ for any function F of the configuration.

Any attempt to evaluate the integrals in (7.53) in a straightforward way is bound to fail. Suppose one were to try approximating them as finite sums: breaking up the multidimensional domain of integration into a very large number of small sub-domains, evaluating the integrand at some interior point in each such sub-region and multiplying by the volume of that sub-region, and then summing the results. The problem with that is that the values of the integrands at all or virtually all the points at which they were evaluated would be 0 or near 0; practically the whole of both integrals in (7.53) comes from only a minute sub-space of the whole $3N$-dimensional configuration space. That is because at the density N/V of a liquid practically every point in the multidimensional integration volume corresponds to a configuration in which at least two – usually many – of the molecules are so close together that the interaction energy W is positive and enormously greater than kT, so that exp $(-W/kT)$ is virtually nil. If the intermolecular potential $\phi(r)$ had an infinitely hard core, W would be strictly ∞ and exp $(-W/kT)$ strictly 0 in practically every configuration.

The only practical way to evaluate or approximate the integrals in (7.53) is to concentrate the effort on just those regions of configuration space in which the Boltzmann factor is not too small; i.e., on those configurations that would be reasonably probable – would occur with reasonable frequency – in the equilibrium system. Instead of thinking of the integrands in the integrals in (7.53) as $F \exp(-W/kT)$ and $\exp(-W/kT)$, and evaluating them at points chosen uniformly, without bias, throughout the domain of integration – giving equal weight to equal elements of volume $d\tau_1 \cdots d\tau_N$ in the multidimensional space – we should think of the integrands as F and 1, respectively, but then choose the points at which to evaluate them with a probability, or frequency, proportional to $\exp(-W/kT)$. In principle the two procedures are equivalent, but in practice the difference is that between certain failure and probable success. Choosing the region of configuration space on which to concentrate one's effort according to the values of $\exp(-W/kT)$ in the region – more generally, to concentrate one's effort where the integrand is large and not waste resources evaluating it where it contributes negligibly to the integral – is an example of *importance sampling*.

In the present context, one would choose some large but manageable number n of configurations – say $n = 10^5$ or 10^6 – choosing them with a probability, or weight, proportional to $\exp(-W/kT)$. Let $\mathbf{r}_i^{(m)}$ be the location of molecule

i in the mth such configuration. Then the resulting estimate of $\bar{F}$ from (7.53) is

$$\bar{F} \simeq \frac{\sum\limits_{m=1}^{n} F\left(\mathbf{r}_1^{(m)}, \ldots, \mathbf{r}_N^{(m)}\right)}{\sum\limits_{m=1}^{n} 1}$$

$$= \frac{1}{n} \sum_{m=1}^{n} F\left(\mathbf{r}_1^{(m)}, \ldots, \mathbf{r}_N^{(m)}\right). \tag{7.54}$$

Since the n configurations m are chosen with the Boltzmann weight $\exp(-W/kT)$, the $\bar{F}$ in (7.54) is the Boltzmann-weighted average of F (the sum of its values in the n configurations, divided by n), just as is the $\bar{F}$ in (7.53).

Even with importance sampling the number of configurations n that must be sampled to achieve reasonable accuracy at a given density N/V becomes rapidly greater as N and V increase. In practice one applies the method only to systems of about the same size as those simulated by molecular dynamics; N equal to a few hundred or a thousand is typical. To minimize finite-size effects one would again impose periodic boundary conditions, as in Fig. 7.4.

There are many ways of insuring that the configuration space is sampled with the Boltzmann weight $\exp(-W/kT)$. Here is one of the most popular, called the Metropolis algorithm. (Naturally, one programs a computer to do all this.) Choose some reasonably low-energy starting configuration – the molecules positioned more or less uniformly in the volume V, say, without excessive penetration of the repulsive cores of their potentials. Then choose one of the molecules at random and let it move a little bit in some randomly chosen direction. (In a variant of this scheme one allows each of the molecules in turn to make such a move, in regularly repeated sequence.) Evaluate W in both the old and the new configurations. If W in the new configuration, W_{new}, is less than W in the old, W_{old}, the move is accepted and the new configuration becomes the starting one for the next move. If, instead, $W_{\text{new}} > W_{\text{old}}$, the new one is accepted only provisionally: a random number s in the interval $0 \leq s \leq 1$ is generated, and if $\exp[-(W_{\text{new}} - W_{\text{old}})/kT] > s$ the move is accepted, while if $\exp[-(W_{\text{new}} - W_{\text{old}})/kT] < s$ the move is rejected and the starting configuration remains the starting one for the next move. This criterion becomes especially simple for hard spheres that do not otherwise interact, for which $W = \infty$ when any spheres interpenetrate and $W = 0$ otherwise. Then a move is accepted if as a result of it no two spheres have been made to overlap; otherwise, it is rejected. In the general case it is essential that the random number generator generate values of s uniformly in the interval $0 \leq s \leq 1$; i.e., without bias. The

7.4 Computer simulation: Monte Carlo

effects of even a slight bias can be magnified by a long run. As long as s can be anywhere in the interval $0 \leq s \leq 1$ with equal probability, the foregoing acceptance criterion for a move with $W_{\text{new}} > W_{\text{old}}$ means it is accepted with probability $\exp[-(W_{\text{new}} - W_{\text{old}})/kT]$. Note that this $\to 1$ as $W_{\text{new}} \to W_{\text{old}}$.

If moves are accepted too often it means one has been too timid and has made the step size too small; it will take too long to get a good sampling of the important parts of the configuration space. The method is equally inefficient if too large a fraction of the moves are rejected, as will surely be the case if the step size is too large: a large step is almost certain to produce a configuration in which there are molecules that are strongly repelling, so that $\exp[-(W_{\text{new}} - W_{\text{old}})/kT] \approx 0$. One usually chooses a step size such that roughly half the moves are accepted and half rejected. It is the element of chance in accepting or rejecting a move that gives the Monte Carlo method its name.

The process just described generates a chain of configurations: a chain of points in the configuration space. It remains now to show that these configurations have been generated with the proper Boltzmann weight, proportional to $\exp(-W/kT)$. It is convenient to think of there being only finitely many possible configurations, say some huge but finite number ν, and labeled by a discrete index m or ℓ running from 1 to ν. Think of each Monte Carlo step as taking a certain "time," and ask, what is the ratio of the "rates" of the transitions $m \to \ell$ and $\ell \to m$? Suppose $W_\ell > W_m$. Then by the Metropolis algorithm the transition $\ell \to m$, if attempted, would occur with certainty, while $m \to \ell$, if attempted, would occur only with probability $\exp[-(W_\ell - W_m)/kT]$. Therefore the ratio of the transition rates $k_{m \to \ell}$ and $k_{\ell \to m}$ is

$$\frac{k_{m \to \ell}}{k_{\ell \to m}} = e^{-(W_\ell - W_m)/kT}. \tag{7.55}$$

But as one knows from chemical kinetics, such a ratio of "rate constants" is the "equilibrium constant"; i.e., it is what, after long times, when equilibrium has been achieved, would be the relative frequency of occurrence of the states ℓ and m. We see from (7.55) that this is just the ratio of the Boltzmann factors $\exp(-W_\ell/kT)$ and $\exp(-W_m/kT)$, as required.

This shows us also that the configurations generated in a Monte Carlo chain, if the chain is long enough and if we ignore the earliest configurations so generated (say, the first 50,000 or 100,000), are typical equilibrium configurations of the model fluid. The Monte Carlo method is silent about the velocities of the molecules – the rules by which the configuration evolves are not the laws of Newtonian dynamics – but the configurations it gives rise to are as reliably equilibrium configurations as are those from the method of molecular dynamics, §7.3. We know already that with the choice $F = W$ the $\bar{F}$ obtained with (7.54) from n such configurations is E_intermol and that adding the ideal-gas $E_\text{i.g.}$ to it would then give us $E(N, V, T)$. Also, from such equilibrium configurations, which are stored in the computer, just as from those generated by molecular dynamics, one may evaluate the radial distribution function $g(r)$ by (7.19), and (if W is the sum of pair interactions as in (7.51)) the equation of state $p = p(\rho, T)$ from (7.33) and (7.41) or from (7.43). For hard spheres one uses (7.50).

It is well to recall that the methods described in this section and in §7.3 depend on the availability of efficient, high-speed computing. We have now seen how essential this has been in giving us an accurate picture of liquids and thus making possible a statistical mechanical theory of the liquid state.

8
Quantum ideal gases

8.1 Bose–Einstein and Fermi–Dirac statistics versus Boltzmann statistics

It was remarked in Chapter 2, §2.1, that there are circumstances in which the "Boltzmann statistics" for an ideal gas – the approximation in Eq. (2.6), in which the total partition function is decomposed into a product of single-molecule partition functions, with division by $N_1!\, N_2!\ldots$ to correct for particle identity – is inaccurate even if the gas is still ideal (non-interacting particles). That happens when the number of single-particle states within the energy kT of the ground state is no longer very much greater than the number of particles. (Recall Fig. 2.1 and the accompanying discussion.)

We noted that when the Boltzmann statistics is no longer an adequate approximation one must distinguish the particles as fermions, no two of which can be in the same single-particle state (counting states of different spin as different), or as bosons, for which there is no restriction on multiple occupancy of the single-particle states. Fermions are particles of half-odd-integer spin quantum number; they could be composite particles consisting of an odd number of spin-$\frac{1}{2}$ particles. Protons, neutrons, and electrons, all of which are of spin $\frac{1}{2}$, are fermions. Bosons are particles of integer spin; they could be spinless, or of spin 1, etc., or could be composites consisting of an even number of fermions. The nuclei of helium atoms of mass number 4 consist of two protons and two neutrons, and the neutral atom has two electrons outside the nucleus; ^{4}He atoms are bosons.

We also noted in §2.1 that the Boltzmann statistics would break down at high densities or at low temperatures (even if the gas were still ideal under those conditions), because that is when multiple occupancy of the single-particle states would be a serious possibility. It was remarked that an electron "gas" at the density of the valence electrons in a metal is such a case; that is so even to extremely high temperatures, as we shall see. Another such case is that of ^{4}He at the density and at the very low temperatures – just a few kelvin – of the liquid. These are hardly ideal gases – the interparticle interactions are very strong – but the corresponding ideal gases at those densities and temperatures already show many of the quantum effects seen in real metals and in liquid helium, and so, especially because of their relative simplicity, the quantum ideal gases are still valuable models.

The appropriate replacement for Eq. (2.6) for an ideal gas of fermions is called the Fermi–Dirac statistics; for bosons, the Bose–Einstein statistics.

We call a system an ideal gas when its constituent molecules do not interact with one another. In the Boltzmann statistics, as we saw in Chapter 2, Eq. (2.18), the equation of state of such a system is the "ideal-gas" law, $pV = NkT$, which is also said to be the equation of state of a "perfect" gas. But, as we shall see, when the Boltzmann statistics is no longer a good approximation this is no longer the equation of state, even when the gas is ideal in the sense of its molecules' not interacting with each other.

For any system, ideal gas or not, the partition function Z is as given in Chapter 1, Eq. (1.12),

$$Z = \sum_i e^{-E_i/kT}, \tag{8.1}$$

where the summation is over the quantum states i of the system as a whole, each of energy E_i. Suppose now that the system consists of N particles, all of the same kind, and that they do not interact with each other, so that the system is in that sense an ideal gas. Let r index the quantum states of the single particles. Then a state i of the whole system is specified by specifying the number, n_r, of particles in each single-particle state r:

$$i : \{n_0, n_1, n_2, \ldots, n_r, \ldots\}. \tag{8.2}$$

In Fig. 8.1 is a hypothetical energy-level diagram showing the energies ε_r of the single-particle states $r = 0, 1, 2, \ldots$, and indicating the "occupation numbers" n_r as a number of particles (indicated by filled circles) in each state. In the example in the figure $n_0 = 2, n_1 = 0, n_2 = 4, n_3 = 1, n_4 = 2, \cdots$. This is one state i of the whole system. Note that it is only the *number* of particles in each state r, not their identity, that one specifies in determining the overall state i of the system. Since the particles are identical, permuting them while leaving the number n_r in each single-particle state r unchanged leaves the state i of the system unchanged.

The difference between fermions and bosons is that for fermions each n_r may be only 0 or 1 while for bosons an n_r may be any non-negative integer $0, 1, 2, 3, \cdots$. In the example in Fig. 8.1 the particles must be bosons.

8.1 BE and FD vs. Boltzmann statistics

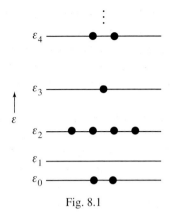

Fig. 8.1

The occupation numbers n_r are of states, not of energy levels. If, for example, the energies ε_3 and ε_4 in Fig. 8.1 happened to be equal, the number of particles occupying that *level* would be $n_3 + n_4 = 3$, but $r = 3$ and $r = 4$ would still be two different single-particle states with occupation numbers $n_3 = 1$ and $n_4 = 2$. [In the familiar picture of the energy levels of an electron in a central-force field, each s level in the analog of Fig. 8.1 would contain two states (the energy of an isolated electron being independent of whether it is spin ↑ or spin ↓, but with these now counting as two different states), each p level would encompass six states, etc.]

With ε_r the energy in the single-particle state r, the total energy E_i in the system state i that is specified by the set of occupation numbers $\{n_r\}$ is

$$E_i = n_0\varepsilon_0 + n_1\varepsilon_1 + n_2\varepsilon_2 + \cdots = \sum_r n_r \varepsilon_r. \tag{8.3}$$

At the same time the sum over states i in (8.1) means a sum over all possible values of all the occupation numbers $n_0, n_1, n_2, \ldots$, remembering that for fermions the only possible value of each is 0 or 1 while for bosons each n_r may be any non-negative integer. Both for fermions and for bosons, however, there is here an additional restriction on the possible values of the occupation numbers: that whatever the state i, that is, whatever may be the values of the individual occupation numbers n_r, the sum of the n_r over all states r must be the total number of particles, N, in the system:

$$n_0 + n_1 + n_2 + \cdots = \sum_r n_r = N. \tag{8.4}$$

Thus, (8.1), with E_i given by (8.3), is

$$Z = \sum_{n_0} \sum_{n_1} \sum_{n_2} \cdots e^{-\frac{1}{kT}\sum_r n_r \varepsilon_r} \tag{8.5}$$
$$\scriptstyle (n_0 + n_1 + \cdots = N)$$

where the statement in parentheses beneath the multiple sum is a reminder of the constraint (8.4) on the possible values of the summation variables. The multiple sum (8.5) is then over all the distinct sets of allowed values $n_0, n_1, n_2, \ldots$ of the occupation numbers n_r; for fermions an allowed set of occupation numbers is any set of 0's and 1's that add up to N, while for bosons an allowed set of occupation numbers is any set of non-negative integers that add up to N.

Exercise (8.1). Suppose there were only a finite number, say 4, of single-particle states r, so $r = 0, 1, 2, 3$, say, and only 3 particles, $N = 3$. Then what would be the possible sets of occupation numbers n_0, n_1, n_2, n_3 if the particles were fermions, or if they were bosons?

Solution. When the particles are fermions it is easy to enumerate the possible sets of the n_r: the only ways we can have no more than one particle in each of the four states and yet have three particles altogether are to have $n_0, n_1, n_2, n_3 = $ 1, 1, 1, 0 or 1, 1, 0, 1 or 1, 0, 1, 1 or 0, 1, 1, 1. For bosons there are many more possibilities. We could have all three particles in the same state, and that could happen four ways: $n_0, n_1, n_2, n_3 = 3, 0, 0, 0$ or $0, 3, 0, 0$ or $0, 0, 3, 0$ or $0, 0, 0, 3$. We could have two particles in one state and one in another, and that could happen twelve ways: $n_0, n_1, n_2, n_3 = 2, 1, 0, 0; 2, 0, 1, 0; 2, 0, 0, 1; 1, 2, 0, 0; 0, 2, 1, 0; 0, 2, 0, 1; 1, 0, 2, 0; 0, 1, 2, 0; 0, 0, 2, 1; 1, 0, 0, 2; 0, 1, 0, 2; 0, 0, 1, 2$. Finally, we could have one particle in each of three different states, and that could happen in the same four ways as for fermions: $n_0, n_1, n_2, n_3 = $ 1, 1, 1, 0; 1, 1, 0, 1; 1, 0, 1, 1; 0, 1, 1, 1. Thus, if the particles were bosons there would be altogether twenty different allowed sets of occupation numbers n_0, n_1, n_2, n_3, but only four of these twenty would be allowed if the particles were fermions.

The Boltzmann statistics [Eq. (2.6)] makes no distinction between fermions and bosons, and approximates the multiple sum (8.5) by

$$Z = \frac{1}{N!} z^N \quad \text{(Boltzmann statistics)} \qquad (8.6)$$

where z is the single-molecule partition function

$$z = \sum_r e^{-\varepsilon_r / kT}; \qquad (8.7)$$

but we know that this is based on an inexact enumeration of the states of the system. We could obtain Z exactly if we could evaluate the sum in (8.5). As it stands, however, this is a formidable problem. What makes it hard is that the sums over n_0, n_1, etc. are not independent of each other; they are coupled by

8.1 BE and FD vs. Boltzmann statistics

the requirement that the sum of the n_r be N. Had that not been the case – had we been allowed to sum over each of the n_r independently of the others – we could easily have evaluated the multiple sum: the summand, being itself the exponential of a sum, is the product of exponentials, each of which depends on only one of the summation variables n_r, so the multiple sum would then just have been a product of independent sums. But, alas, the sums over the various n_r in (8.5) *are* coupled by the requirement $\sum n_r = N$ and so are *not* independent; the multiple sum is *not* the product

$$\sum_{n_0}\left(e^{-\varepsilon_0/kT}\right)^{n_0} \sum_{n_1}\left(e^{-\varepsilon_1/kT}\right)^{n_1}\cdots \tag{8.8}$$

of independent sums.

Exercise (8.2). Compare the two double sums

$$S_1 = \sum_{m=0}^{\infty}\sum_{n=0}^{\infty} x^m y^n$$

and

$$S_2 = \sum_{\substack{m=0\ n=0 \\ (m+n=6)}}^{\infty}\sum x^m y^n,$$

where in S_1 the summations over m and n are independent while in S_2 they are not: in each term of the double sum S_2 the sum of m and n is required to be 6.

Solution.

$$S_1 = \left(\sum_{m=0}^{\infty} x^m\right)\left(\sum_{n=0}^{\infty} y^n\right)$$

$$= \frac{1}{1-x}\cdot\frac{1}{1-y}.$$

S_2, by contrast, is the sum of $x^m y^n$ over only those pairs of non-negative integers m, n that add up to 6. Those pairs are $m, n = 0, 6;\ 1, 5;\ 2, 4;\ 3, 3;\ 4, 2;\ 5, 1;$ and $6, 0$. Thus,

$$S_2 = y^6 + xy^5 + x^2y^4 + x^3y^3 + x^4y^2 + x^5y + x^6,$$

which may be rearranged to

$$S_2 = (y^7 - x^7)/(y - x).$$

(To verify that this form of S_2 is the same as the preceding, multiply the preceding one by $y - x$.) Note that S_2, unlike S_1, is not decomposable into the product of a function of x alone and a function of y alone. That is because the constraint $m + n = 6$ in the double sum S_2 couples x and y, even though x and the summation index m, and y and the summation index n, appear in separate factors of the summand. Nevertheless, all the terms in S_2 are present in S_1: if S_1 is written out as the product $(1 + x + x^2 + \cdots)(1 + y + y^2 + \cdots)$, which is then expanded, it is seen that all the terms $y^6 + xy^5 + \cdots$ of S_2 appear among the terms in that expansion, as a small subset of them.

The problem of evaluating the multiple sum (8.5) subject to the constraint $n_0 + n_1 + \cdots = N$ – and expressing the result as an asymptotic formula holding in the asymptotic limit of very large N, which is all that would be of interest – is not insurmountable, but a better strategy is to avoid the problem entirely by evaluating a different partition function that is not subject to that constraint. The appropriate partition function, from which the constraint of fixed N has been removed and in which N has been replaced as an independent variable by a different thermodynamic variable, is the *grand-canonical* (or, more briefly, the *grand*) partition function.

In §8.2 we derive this new partition function from the canonical partition function $Z(T, V, N)$ and we learn what its properties are. Then in §8.3 we evaluate it for the quantum ideal gases. That calculation starts with the formula (8.5) for Z, which we have here left unevaluated.

8.2 The grand-canonical partition function

To construct the new partition function we take as a model Eq. (1.21) of Chapter 1, which is a transformation from the microcanonical $W(E, V, N)$, in which all the independent variables are extensive, to the canonical $Z(T, V, N)$, in which the intensive T replaces the extensive E as an independent variable. That transformation was effected by multiplying the partition function that is to be transformed, $W(E)$, by the Boltzmann factor $\exp(-E/kT)$, and then integrating over all possible E. To follow that model, then, we must multiply $Z(T, V, N)$ by the appropriate analog of the Boltzmann factor – containing N and the intensive variable that will replace N as an independent variable in the new partition function – and then sum over all possible values of N.

To see what must play the role of the Boltzmann factor in this transformation, note that of the various thermodynamic potentials it is only $S(E, V, N)$ (now using E rather than U for the energy) that has all extensive variables

8.2 The grand-canonical partition function

as its arguments. In a multicomponent system, more generally, this would be $S(E, V, N_1, N_2, \ldots)$ where N_i is the number of molecules of species i in the mixture, but for our purposes we need consider only a one-component system. The differential thermodynamic identity

$$dS = \frac{1}{T}dE + \frac{p}{T}dV - \frac{\mu}{T}dN \qquad (8.9)$$

shows how S varies as a function of each of those arguments: $(\partial S/\partial E)_{V,N} = 1/T$, etc. With each extensive variable X there is then associated an intensive variable,* which is the coefficient of dX in (8.9); thus, $1/T$ with E, p/T with V, and $-\mu/T$ with N.

Then just as the pair E and $1/T$ are associated in the Boltzmann factor $\exp(-E/kT)$, so now, in the new transformation we contemplate making, the pair N and $-\mu/T$ will be associated in a factor $\exp(\mu N/kT)$, and this will play the role that $\exp(-E/kT)$ played in the earlier transformation (1.21). The new partition function, which we call Ξ, is then

$$\Xi(T, V, \mu) = \sum_{N=0}^{\infty} e^{\mu N/kT} Z(T, V, N), \qquad (8.10)$$

a function of T, V, and the chemical potential μ. This is the grand (-canonical) partition function.

Referring again to the earlier transformation (1.21), recall that the integrand $W(E)\exp(-E/kT)$ is proportional to the probability distribution $Q(E)$ of the energy, which is such that $Q(E)dE$ is the probability that the energy of the system of given T, V, and N will be found in the infinitesimal range of energies E to $E + dE$. The distribution $Q(E)$ as a function of E was pictured in Fig. 1.4. Since it differs only by a normalization constant (viz., the inverse of the partition function, Z^{-1}; see Eq. (1.22)) from $W(E)\exp(-E/kT)$, the same figure also depicts the E dependence of the latter. This is shown here in Fig. 8.2(a). The summand $Z(N)\exp(\mu N/kT)$ in (8.10) is sketched in Fig. 8.2(b) and has an analogous interpretation: it is proportional to the probability that the number of particles in the system of given T, V, and μ will be found to be N. When normalized by division by the grand partition function Ξ, it is the probability distribution itself.

---

* These intensive variables belong to a class of thermodynamic variables called "fields," which have the property of having uniform values throughout a system that is in thermodynamic equilibrium – even when the system consists of two or more phases and so is not itself spatially uniform. There are other intensive variables, like the density, that do not have this property. The name "field" for a variable that does have it was introduced into thermodynamics by Griffiths and Wheeler, in 1970 (*Phys. Rev. A* **2**, 1047). It comes from the thermodynamics of magnetic systems, in which one of the field variables is the applied magnetic field, which, divided by T, is the intensive conjugate of the extensive magnetization.

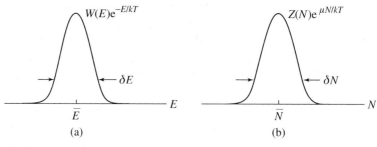

Fig. 8.2

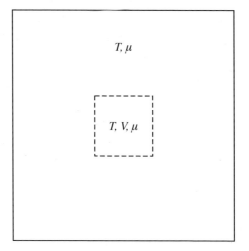

Fig. 8.3

It was remarked in Chapter 1 that putting the system in intimate thermal contact with a thermostat to fix its temperature at T subjects it to fluctuations in its energy. It is just those energy fluctuations that are described by the probability distribution in Fig. 1.4, and now in Fig. 8.2(a). Similarly, when it is the chemical potential μ that is prescribed instead of the number of particles N, that number may fluctuate. Such a system of given T, V, and μ may be thought of as a macroscopically large sub-system of an even much larger (in principle, infinitely large) system at that T and μ, as in Fig. 8.3. The sub-system, which is the system of interest, is that outlined by the dashed lines in the figure. Those lines do not indicate walls; they merely show what sub-region of the larger enclosure it is that constitutes our system. Molecules move freely into and out of this sub-volume from and to the larger volume of which it is a part. Because of this free exchange of molecules between the

8.2 The grand-canonical partition function

system of interest and its surroundings the chemical potential μ of the system is fixed and equal to that in the larger system; but that is also why the number of molecules N that are in the sub-volume V at any one moment is subject to fluctuation about some mean number, $\bar{N}$. It is these fluctuations in N at fixed T, V, and μ that are described by the probability distribution in Fig. 8.2(b).

The widths of the energy and particle-number probability distributions, called δE and δN in Fig. 8.2, give the magnitudes of the typical fluctuations in E (at fixed T, V, and N) and in N (at fixed T, V, and μ). The fluctuations are about the mean energy $\bar{E}$ and about the mean particle number $\bar{N}$. Fluctuations much greater in magnitude than the widths δE and δN are highly improbable. It was remarked in Chapter 1 that while the extensive $\bar{E}$ is proportional to the size of the system, the width δE is very much smaller, of the order only of the square-root of the system size. Likewise here, in Fig. 8.2(b), the width δN is only of order $\sqrt{\bar{N}}$, much less than $\bar{N}$ itself. On the scale of the averages $\bar{E}$ and $\bar{N}$, then, the typical fluctuations δE and δN are minute: E and N are very narrowly distributed about their means.

This has two important consequences. First, the means $\bar{E}$ and $\bar{N}$ may be identified with the most probable E and N (as already anticipated in Fig. 8.2); i.e., with the E and N at which the respective probability distributions have their maxima. Second, since the fluctuations are too small to be macroscopically discernible, the $\bar{E}$ and $\bar{N}$ at which the distributions peak may be identified as the thermodynamic E at the given T, V, and N, and the thermodynamic N at the given T, V, and μ, respectively. (We already made the identification $\bar{E} = U$ in Chapter 1.) It is this observation that will now allow us to identify the thermodynamic potential, a function of T, V, and μ, that is k times the logarithm of the grand-canonical Ξ, just as the thermodynamic potential $-A(T, V, N)/T$ is k times the logarithm of the canonical Z [Eq. (1.16)] and $S(E, V, N)$ is k times the logarithm of the microcanonical W [Eq. (1.26)]. Although Ξ, Z, and W depend on different sets of independent variables, these three partition functions contain the same information about the thermodynamic properties of the system.

In the discussion of Eq. (1.21) and Fig. 1.4 in Chapter 1 it was remarked that the integrand in (1.21) is so strongly peaked about its maximum at $E = \bar{E}$ that the whole integral is just the integrand evaluated at $\bar{E}$, multiplied by some ΔE that is of the order of the width δE of the peak in the energy distribution. That is the content of Eq. (1.23).

Now, analogously, because the particle-number distribution is so strongly peaked about its maximum at $N = \bar{N}$, the whole sum in (8.10) is just the summand itself evaluated at $N = \bar{N}$, multiplied by some ΔN of order δN. The logarithm of ΔN is only of order $\ln N$, which is sub-extensive. Therefore, taking the logarithms of both sides of (8.10), and neglecting sub-extensive corrections,

$$\ln \Xi(T, V, \mu) = \mu \bar{N}/kT + \ln Z(T, V, \bar{N})$$
$$= \mu \bar{N}/kT - A(T, V, \bar{N})/kT \qquad (8.11)$$

with $\bar{N}$ the thermodynamic N at the given T, V, and μ. But with $\bar{N}$ the thermodynamic N, the product $\mu \bar{N}$ is the Gibbs free energy G, which – as one learns from their definitions in thermodynamics – differs from the Helmholtz free energy A by the pressure–volume product, pV. Therefore (8.11) is

$$pV = kT \ln \Xi, \qquad (8.12)$$

identifying the logarithm of the grand partition function as pV/kT.

Equation (8.12) also tells us that at fixed T and μ the logarithm of $\Xi(T, V, \mu)$ will prove to be directly proportional to V. That is because all the system's intensive functions, including p, are fixed by T and μ alone; for fixed T and μ they are independent of the size of the system. Thus, once the grand partition function $\Xi(T, V, \mu)$ has been evaluated, removing the factor V from its logarithm and multiplying the resulting function by kT yields the thermodynamic function $p(T, \mu)$.

This $p(T, \mu)$ is the thermodynamic potential appropriate to the independent variables T and μ; it allows all the intensive properties of the system to be obtained by differentiation. That is the essential content of the Gibbs–Duhem equation of thermodynamics, which for a one-component system is just

$$N d\mu = -S dT + V dp. \qquad (8.13)$$

Rearranged, this is

$$dp = \frac{N}{V} d\mu + \frac{S}{V} dT. \qquad (8.14)$$

The differential coefficients are the number density $\rho = N/V$ and the entropy density $s = S/V$:

$$\rho = \left(\frac{\partial p}{\partial \mu}\right)_T \qquad (8.15)$$

$$s = \left(\frac{\partial p}{\partial T}\right)_\mu. \qquad (8.16)$$

One may then also obtain the energy density, $u = U/V$, making use of the

8.3 Grand partition function of the quantum ideal gases

thermodynamic relations $N\mu = G = A + pV = U - TS + pV$; thus,

$$u = \rho\mu + Ts - p = \rho\mu - \left(\frac{\partial p/T}{\partial 1/T}\right)_\mu, \quad (8.17)$$

where the second equality may be verified by carrying out the differentiation and then referring to (8.16).

In summary, (8.10) tells us how to construct the grand partition function Ξ for a system of given chemical potential μ starting from the canonical partition function Z for a system of a given number of particles N; (8.12) tells us how to obtain the pressure $p(T, \mu)$ from Ξ; and (8.15)–(8.17) tell us how then to obtain the number, entropy, and energy densities, $\rho(T, \mu)$, $s(T, \mu)$, and $u(T, \mu)$, from $p(T, \mu)$. With these tools, we may now resume our attack on the problem of the quantum ideal gases.

8.3 Grand partition function of the quantum ideal gases

We now return to the partition function Z of the quantum ideal gases, which we left unevaluated in (8.5).

According to (8.10), to form the grand partition function $\Xi(T, V, \mu)$ from the Z in (8.5) we must multiply that Z by $\exp(\mu N/kT)$ and then sum over all N from 0 to ∞. Since the terms in the original sum (8.5) are subject to the restriction $n_0 + n_1 + \cdots = N$, multiplying by $\exp(\mu N/kT)$ preparatory to summing over N is the same as multiplying by $\exp[(\mu/kT)\sum_r n_r]$; i.e.,

$$e^{\mu N/kT} Z = \sum_{\substack{n_0 \ n_1 \ n_2 \\ (n_0+n_1+\cdots=N)}} \cdots e^{-\frac{1}{kT}\sum_r n_r(\varepsilon_r - \mu)}; \quad (8.18)$$

and it is this that we wish now to sum over all N from 0 to ∞. That would mean summing over all possible values of the occupation numbers $n_0, n_1, \ldots$ subject to the restriction that $n_0 + n_1 + \cdots = N$ and then summing over all N. But that is exactly the same as just summing over all possible values of the $n_0, n_1, \ldots$ with *no* restriction on their total. [Cf. Ex. (8.3), below.] Therefore the grand partition function Ξ is

$$\Xi = \sum_{N=0}^\infty e^{\mu N/kT} Z$$

$$= \sum_{n_0} \sum_{n_1} \sum_{n_2} \cdots e^{-\frac{1}{kT}\sum_r n_r(\varepsilon_r - \mu)}. \quad (8.19)$$

For fermions each summation variable n_r now takes the values 0 and 1 independently of the others, while for bosons each n_r takes all non-negative integer values 0, 1, 2, 3, ... independently of the others; there is no longer any restriction on the value of $n_0 + n_1 + n_2 + \cdots$ in the terms of the multiple sum. Thus, by simply subtracting a fixed μ from each ε_r in (8.5), and lifting the restriction on the possible values of the sum of the summation variables, we have transformed the canonical partition function $Z(T, V, N)$ into the grand partition function $\Xi(T, V, \mu)$.

Exercise (8.3). Show by explicitly evaluating

$$S_3 = \sum_{N=0}^{\infty} \left[\sum_{\substack{m=0 \\ (m+n=N)}}^{\infty} \sum_{n=0}^{\infty} x^m y^n \right]$$

that it is identical to the

$$S_1 = \sum_{m=0}^{\infty} \sum_{n=0}^{\infty} x^m y^n$$

$$= \frac{1}{1-x} \cdot \frac{1}{1-y}$$

of Ex. (8.2).

Solution. The quantity in square brackets in S_3 is

$$(y^{N+1} - x^{N+1})/(y - x)$$

[cf. the evaluation of S_2 in Ex. (8.2)], so

$$S_3 = \frac{1}{y-x}\left(y \sum_{N=0}^{\infty} y^N - x \sum_{N=0}^{\infty} x^N \right)$$

$$= \frac{1}{y-x}\left(\frac{y}{1-y} - \frac{x}{1-x} \right)$$

$$= \frac{1}{y-x} \frac{y-x}{(1-y)(1-x)}$$

$$= \frac{1}{(1-y)(1-x)},$$

which is identical to S_1.

We note that the multiple sum in (8.19), unlike that in (8.5), consists of independent sums over the independent summation variables $n_0, n_1, \ldots$, with a

8.3 Grand partition function of the quantum ideal gases

summand that is a product of factors,

$$e^{-n_0(\varepsilon_0-\mu)/kT} e^{-n_1(\varepsilon_1-\mu)/kT} \cdots, \qquad (8.20)$$

each depending on only one of the summation variables. The multiple sum (8.19), therefore, unlike (8.5), *does* decompose into a product of independent sums,

$$\Xi = \sum_{n_0} e^{-n_0(\varepsilon_0-\mu)/kT} \sum_{n_1} e^{-n_1(\varepsilon_1-\mu)/kT} \cdots$$

$$= \prod_r \sum_n e^{-n(\varepsilon_r-\mu)/kT}. \qquad (8.21)$$

The subscript r has been dropped from the summation indices n_r; each is a dummy variable of summation and so can be called anything; and since the sums are now independent, each n_r in turn can simply be called n – or anything else. The subscript r on the ε_r is crucial, however; in one sum of the infinite product of sums ε_r is ε_0, in another it is ε_1, etc. For fermions, we recall, each sum over n in (8.21) contains only the two terms $n = 0$ and $n = 1$, while for bosons each is an infinite series, with n ranging from 0 to ∞.

For fermions, then,

$$\sum_n e^{-n(\varepsilon_r-\mu)/kT} = \sum_{n=0}^{1} \left[e^{-(\varepsilon_r-\mu)/kT}\right]^n$$

$$= 1 + e^{-(\varepsilon_r-\mu)/kT} \quad \text{(fermions)}, \qquad (8.22)$$

while for bosons,

$$\sum_n e^{-n(\varepsilon_r-\mu)/kT} = \sum_{n=0}^{\infty} \left[e^{-(\varepsilon_r-\mu)/kT}\right]^n$$

$$= \frac{1}{1 - e^{-(\varepsilon_r-\mu)/kT}} \quad \text{(bosons)}. \qquad (8.23)$$

Therefore, from (8.21), the grand partition function is

$$\Xi = \prod_r \left[1 \pm e^{-(\varepsilon_r-\mu)/kT}\right]^{\pm 1}$$

(top sign for fermions, bottom sign for bosons). (8.24)

These are said to be the Fermi–Dirac statistics (for fermions) and the Bose–Einstein statistics (for bosons). They contrast with the Boltzmann statistics, for which the canonical partition function Z is given by (8.6) with (8.7). To make a proper comparison with (8.24) we should evaluate the grand partition function Ξ of the ideal gas in the Boltzmann statistics. From (8.6) and (8.10) this is

$$\Xi = \sum_{N=0}^{\infty} \frac{1}{N!} z^N e^{\mu N/kT}$$

$$= \sum_{N=0}^{\infty} \frac{1}{N!} \left(z e^{\mu/kT}\right)^N$$

$$= e^{z \exp(\mu/kT)} \quad \text{(Boltzmann statistics)}. \tag{8.25}$$

(Recall the expansion $e^x = \sum_{n=0}^{\infty} \frac{1}{n!} x^n$.) From the equation (8.7) for the single-particle partition function z this may be stated more explicitly as

$$\Xi = e^{\left[\sum_r \exp(-\varepsilon_r/kT)\right] \exp(\mu/kT)}$$

$$= e^{\sum_r \exp[-(\varepsilon_r - \mu)/kT]}$$

$$= \prod_{r=0}^{\infty} e^{e^{-(\varepsilon_r - \mu)/kT}} \quad \text{(Boltzmann statistics)}, \tag{8.26}$$

(note the exponentials of exponentials !), in which form it may be compared directly with the corresponding formula (8.24) in the Fermi–Dirac and Bose–Einstein statistics.

Suppose it were the case that the single-particle ground-state energy level ε_0, the chemical potential μ, and the temperature T, were such that

$$e^{-(\varepsilon_0 - \mu)/kT} \ll 1. \tag{8.27}$$

(There is an arbitrary additive constant in the chemical potential μ arising from the arbitrary 0 of energy but there is no arbitrariness in the difference $\varepsilon_0 - \mu$.) Then the same inequality would hold even more strongly for all the other $\exp[-(\varepsilon_r - \mu)/kT]$ because $\varepsilon_r \geq \varepsilon_0$; so it would follow from (8.27) that

$$e^{e^{-(\varepsilon_r - \mu)/kT}} \simeq 1 + e^{-(\varepsilon_r - \mu)/kT} \tag{8.28}$$

for all r (because $\exp x \sim 1 + x$ for small x), and also that

$$\left[1 - e^{-(\varepsilon_r - \mu)/kT}\right]^{-1} \simeq 1 + e^{-(\varepsilon_r - \mu)/kT} \tag{8.29}$$

for all r (because $1/(1-x) \sim 1 + x$ for small x). But then all three formulas for Ξ – (8.24) with the top and bottom signs in the Fermi-Dirac and Bose–Einstein statistics, respectively, and (8.26) in the Boltzmann statistics – would be identical. We therefore see that (8.27) is the condition for the validity of

8.3 Grand partition function of the quantum ideal gases

the Boltzmann statistics for a gas of non-interacting particles (ideal gas); and that when it holds, the distinctions among the Fermi–Dirac, Bose–Einstein, and Boltzmann statistics disappear.

What is the meaning of the condition (8.27) for the validity of the Boltzmann statistics? We had already seen in Chapter 1, and recalled at the start of this chapter, that the Boltzmann statistics would be valid – and distinguishing fermions and bosons would be unnecessary – if the number of single-particle states within the energy kT of the ground state were much greater than the number of particles, for then the question of multiple occupancy of the states would cease being an issue. Let us then calculate the mean occupation number $\bar{n}_r$ – the average number of particles occupying the single-particle state r of energy ε_r – in a thermodynamic state of prescribed T and μ, in the Fermi–Dirac, the Bose–Einstein, and the Boltzmann statistics.

It was remarked earlier that the summand in the equation (8.10) for the grand partition function Ξ is proportional to the probability that the system of given T, V, and μ will contain N molecules. The factor Z in that summand is in turn the sum of Boltzmann factors $\exp(-E_i/kT)$, summed over all states i of a system of given N and V [Eq. (1.12)]; and that Boltzmann factor, we know [Eq. (1.8)], is itself proportional to the probability that the system of N molecules in volume V at temperature T will be found in the quantum state i. That summation over all i to give Z, followed by the summation of $Z \exp(\mu N/kT)$ over all N, is what gave us the formula (8.19) for the grand partition function of the quantum ideal gases. The summand in the multiple sum in (8.19) is then proportional to the probability that the system of given T, V, and μ contains $N(= n_0 + n_1 + \cdots)$ particles *and* that the state of that system of $N(= n_0 + n_1 + \cdots)$ particles is that in which n_0 of them are in the single-particle state $r = 0$, and n_1 of them are in the single-particle state $r = 1$, etc. Thus, the probability $P_{n_0, n_1, \ldots}$ that the quantum ideal gas of given T, V, and μ will be in the state in which the occupation numbers are some given allowed set $n_0, n_1, \ldots$ is

$$P_{n_0, n_1, \ldots} = \Xi^{-1} e^{-\frac{1}{kT} \sum_r n_r (\varepsilon_r - \mu)} \quad (8.30)$$

$$= \Xi^{-1} \prod_r e^{-n_r (\varepsilon_r - \mu)/kT}. \quad (8.31)$$

The factor Ξ^{-1} in (8.30) and (8.31) is the required normalization factor; by (8.19), it guarantees that the sum of $P_{n_0, n_1, \ldots}$ over all the allowed $n_0, n_1, \ldots$ is 1.

But (8.31) shows this probability $P_{n_0, n_1, \ldots}$ to be itself the product of factors each depending on only one of the occupation numbers n_r; that is, this is the probability of independent events: the probability that state r is occupied by

n_r particles is independent of the extent of occupancy of any of the other states. That is what going from a system of given N (canonical) to a system of given μ (grand-canonical) accomplished: when N is prescribed instead of μ the occupancies of the different single-particle states are *not* independent, because of the constraint $n_0 + n_1 + \cdots = N$. Thus, here, with T and μ prescribed, the probability $p_r(n_r)$ that the single-particle state r of energy ε_r will be occupied by n_r particles is proportional to $\exp[-n_r(\varepsilon_r - \mu)/kT]$; and, normalized, this probability is

$$p_r(n_r) = \frac{e^{-n_r(\varepsilon_r - \mu)/kT}}{\sum_n e^{-n(\varepsilon_r - \mu)/kT}}, \tag{8.32}$$

where the summation in the denominator is over all the allowed values of n: $n = 0, 1$ for fermions (Fermi–Dirac statistics) and $n = 0, 1, 2, 3, \ldots$ for bosons (Bose–Einstein statistics). The summation index n in the denominator is a dummy variable and could have been called anything, including n_r; the only dependence of $p_r(n_r)$ on n_r comes from the n_r in the exponent in the numerator.

The normalization denominator in (8.32) is just the quantity already evaluated for Fermi–Dirac and Bose–Einstein statistics in (8.22) and (8.23), but for our present purpose, which is to evaluate the average, $\bar{n}_r$, it is more convenient to leave $p_r(n_r)$ in the form (8.32). Then that average is

$$\bar{n}_r = \sum_{n_r} n_r p_r(n_r)$$

$$= \frac{\sum_n n\, e^{-n(\varepsilon_r - \mu)/kT}}{\sum_n e^{-n(\varepsilon_r - \mu)/kT}}, \tag{8.33}$$

where now both summation variables n are dummy variables. The average $\bar{n}_r$ depends on the state r only via the energy ε_r.

If we temporarily let

$$x = e^{-(\varepsilon_r - \mu)/kT} \tag{8.34}$$

then (8.33) is

$$\bar{n}_r = \frac{\sum_n n x^n}{\sum_n x^n}$$

$$= \frac{1}{\sum_n x^n} \cdot x \frac{d}{dx} \sum_n x^n$$

$$= x \frac{d}{dx} \ln \sum_n x^n. \tag{8.35}$$

8.3 Grand partition function of the quantum ideal gases

Thus, to evaluate $\bar{n}_r$ we have just to evaluate the sum $\sum_n x^n$ as a function of x, take its logarithm, differentiate with respect to x, and then multiply by x. But [cf. (8.22) and (8.23)]

$$\sum_n x^n = \sum_{n=0}^{1} x^n = 1 + x \quad \text{(Fermi–Dirac)} \tag{8.36}$$

or

$$\sum_n x^n = \sum_{n=0}^{\infty} x^n = \frac{1}{1-x} \quad \text{(Bose–Einstein)}. \tag{8.37}$$

Therefore

$$\bar{n}_r = x \frac{d}{dx} \ln(1+x) = \frac{x}{1+x} = \frac{1}{x^{-1}+1} \quad \text{(Fermi–Dirac)} \tag{8.38}$$

or

$$\bar{n}_r = -x \frac{d}{dx} \ln(1-x) = \frac{x}{1-x} = \frac{1}{x^{-1}-1} \quad \text{(Bose–Einstein)}. \tag{8.39}$$

Then with x as defined by (8.34),

$$\bar{n}_r = \frac{1}{e^{(\varepsilon_r - \mu)/kT} \pm 1} \quad \text{(top sign Fermi–Dirac, bottom sign Bose–Einstein)}. \tag{8.40}$$

This is the mean occupation number of the single-particle state r, of energy ε_r, at given T and μ.

In the Boltzmann statistics of the ideal gas the molecules occupy the single-particle states r independently of each other and there is no restriction on how many may be in each state. The probability that a molecule is in the state r of energy ε_r is proportional to the Boltzmann factor $\exp(-\varepsilon_r/kT)$ [Chapter 1, §1. 2]; when divided by the single-molecule partition function z, given in (8.7), this becomes the normalized probability itself. Thus, on average, the fraction $z^{-1} \exp(-\varepsilon_r/kT)$ of all the molecules are to be found in the state r. Then if the system is at temperature T and contains N particles, the mean occupation number $\bar{n}_r$ is

$$\bar{n}_r = \frac{N}{z} e^{-\varepsilon_r/kT} \quad \text{(Boltzmann statistics)}. \tag{8.41}$$

But it follows from (3.13), which we derived in connection with the theory of chemical equilibrium in dilute gases, that in a one-component system the ratio of the number of molecules N to the single-molecule partition function z is

$$\frac{N}{z} = e^{\mu/kT} \tag{8.42}$$

in the Boltzmann statistics. Then (8.41) becomes

$$\bar{n}_r = e^{-(\varepsilon_r - \mu)/kT} \quad \text{(Boltzmann statistics)}, \tag{8.43}$$

expressed now in terms of μ and T rather than N and T. In this form it is directly comparable with the corresponding formula (8.40) for the mean occupation numbers in the Fermi–Dirac and Bose–Einstein statistics.

On comparing (8.43) with (8.40) we see that when $\exp[(\varepsilon_r - \mu)/kT] \gg 1$, or $\exp[-(\varepsilon_r - \mu)/kT] \ll 1$, the distinction between the Fermi–Dirac and Bose–Einstein statistics disappears and both become the same as the Boltzmann statistics. If that inequality holds for the ground state $r = 0$ then it holds even more strongly for all the other states, since $\varepsilon_r \geq \varepsilon_0$. Thus, (8.27) is again seen to be the condition for validity of the Boltzmann statistics. But when that inequality holds we have $\bar{n}_r \ll 1$, by (8.43) or (8.40); i.e., there is on average very little multiple occupancy of states, even in the Boltzmann or Bose–Einstein statistics, where multiple occupancy is not forbidden. This then confirms the earlier conclusion that the Boltzmann statistics is a good approximation to the statistical mechanics of the ideal gas whenever multiple occupancy of states is a rare event even when not disallowed. That this is the same condition as that the number of single-particle states within kT of the ground state must be much greater than the number of particles will be explicitly verified later.

The comparison of the formulas (8.40) and (8.43) for $\bar{n}_r$ has shown us that the "only" difference between the Fermi–Dirac and Bose–Einstein statistics of the ideal gas, and between them and the Boltzmann statistics, is whether, in the denominator of (8.40), one adds 1 to or subtracts 1 from $\exp[(\varepsilon_r - \mu)/kT]$, or does neither; but those seemingly slight differences, as will be apparent in §§8.4 and 8.5, can have profound consequences.

Before turning to the specific properties of the Fermi–Dirac and Bose–Einstein ideal gases in §§8.4 and 8.5 we must return to the formula (8.24) for their grand partition functions and make it more explicit by incorporating in it explicit expressions for the energy levels ε_r as functions of the volume V. We specialize here to the case of structureless particles, or particles whose internal degrees of freedom may be ignored, perhaps because the temperature is too low for them to be significantly excited. This is certainly an appropriate assumption when the Fermi–Dirac statistics is being applied to electrons or the Bose–Einstein statistics to helium. We will, however, take account of any

8.3 Grand partition function of the quantum ideal gases

possible degeneracy g associated with internal degrees of freedom, such as the spin degeneracy $g = 2$ for electrons.

The single-particle energy levels ε_r are then just the translational energies of a particle in a three-dimensional box, a familiar model in quantum mechanics. Each state r is characterized by a set of three quantum numbers n_x, n_y, n_z, each of which takes all positive integer values,

$$n_x, n_y, n_z = 1, 2, 3, 4, \ldots; \tag{8.44}$$

and if for simplicity we assume a cubical box, of edge length L, then the energy levels ε_r, which we now call $\varepsilon_{n_x,n_y,n_z}$, are

$$\varepsilon_{n_x,n_y,n_z} = \frac{h^2}{8mL^2}\left(n_x^2 + n_y^2 + n_z^2\right) \tag{8.45}$$

where m is the mass of the particle [cf. (2.11)]. These energies are on a scale on which, classically, the minimum energy would be 0, so that the quantum-mechanical zero-point energy is $\varepsilon_{1,1,1} = 3h^2/(8mL^2)$.

Only the macroscopic, large-volume asymptotic limit $L \to \infty$ is of interest, so the relevant energy levels are extremely closely spaced and practically form a continuum. We then ask, how many states, $\omega(\varepsilon)d\varepsilon$, are there with energies between ε and $\varepsilon + d\varepsilon$; i.e., what is the density $\omega(\varepsilon)$ of these single-particle states?

Exactly the same question came up in a different guise in the derivation of the Rayleigh–Jeans law in Chapter 4, §4.2. There we sought the density $G(\nu)$ of vibrational modes of frequency ν in an elastic continuum, with ν^2 determined by the "quantum numbers" n_x, n_y, n_z in the same way that the energy ε is in the present equation (8.45). On comparing (8.45) with (4.6) we see that the connection is

$$\frac{2m}{h^2}\varepsilon \Leftrightarrow \frac{1}{c^2}\nu^2 \tag{8.46}$$

where c in the former problem was the speed of propagation of an elastic wave of frequency ν. In the same macroscopic limit $L^3 = V \to \infty$ as here, we found previously [Eq. (4.8), but now considering only one class of such normal modes, with speed of propagation c] that with $G(\nu)d\nu$ the number of such modes with frequencies in the infinitesimal frequency interval ν to $\nu + d\nu$,

$$G(\nu) = (4\pi V/c^3)\nu^2. \tag{8.47}$$

But $Gd\nu = (c^2/2\nu)Gd(\nu^2/c^2)$, so $(c^2/2\nu)G$ is the density with which ν^2/c^2 is distributed. By (8.47), this density is $2\pi V\nu/c$. Then from the correspondence

(8.46), the density with which $2m\varepsilon/h^2$ is distributed in our present problem is $2\pi V(2m\varepsilon/h^2)^{1/2}$; i.e., the number of states in which $2m\varepsilon/h^2$ lies between $2m\varepsilon/h^2$ and $2m\varepsilon/h^2 + d(2m\varepsilon/h^2)$ is $2\pi V(2m\varepsilon/h^2)^{1/2} d(2m\varepsilon/h^2)$. But the states in which $2m\varepsilon/h^2$ lies in that infinitesimal range are just those in which ε lies in the range ε to $\varepsilon + d\varepsilon$; so the number $\omega(\varepsilon) d\varepsilon$ of the latter, too, is

$$\omega(\varepsilon) d\varepsilon = 2\pi V \sqrt{2m\varepsilon/h^2}\, d(2m\varepsilon/h^2)$$
$$= 2\pi V(2m/h^2)^{3/2} \sqrt{\varepsilon}\, d\varepsilon; \qquad (8.48)$$

i.e., the density $\omega(\varepsilon)$ of such states is

$$\omega(\varepsilon) = 2\pi V(2m/h^2)^{3/2} \sqrt{\varepsilon}. \qquad (8.49)$$

From (8.24), the logarithm of the grand partition function is

$$\ln \Xi = \pm \sum_r \ln\left[1 \pm e^{-(\varepsilon_r - \mu)/kT}\right] \quad \text{(top sign FD, bottom sign BE)}. \qquad (8.50)$$

The sum over the single-particle states r may now, in our large-volume limit, be replaced by an integration over the energies ε, with $\omega(\varepsilon) d\varepsilon$ the number of these states in the interval $d\varepsilon$. We may also include with ω any degeneracy factor g arising from internal states, such as the spin degeneracy $g = 2$ for electrons. Then from (8.49) and (8.50),

$$\ln \Xi = \pm 2\pi g V(2m/h^2)^{3/2} \int_0^\infty \varepsilon^{1/2} \ln\left[1 \pm e^{-(\varepsilon - \mu)/kT}\right] d\varepsilon. \qquad (8.51)$$

The proportionality to V is as we predicted from (8.12). Note that the integration starts from $\varepsilon = 0$. That is because the zero-point energy, being inversely proportional to L^2, vanishes in the thermodynamic limit.

As a convenient abbreviation of symbols, define the "thermal wavelength"

$$\Lambda = \frac{h}{\sqrt{2\pi mkT}}. \qquad (8.52)$$

It is called that because, as we know from the kinetic theory of gases, the average speed with which the molecules move is proportional to $\sqrt{kT/m}$, so the average magnitude of the molecules' momentum is proportional to $\sqrt{mkT}$. Therefore Λ as defined in (8.52) is the mean quantum mechanical (de Broglie) wavelength of the molecules at temperature T. If we now let $x = \varepsilon/kT$ be a new integration variable, then (8.51) with (8.52) becomes

$$\ln \Xi = \pm \frac{2g}{\sqrt{\pi}\Lambda^3} V \int_0^\infty x^{1/2} \ln\left[1 \pm e^{-x+\mu/kT}\right] dx. \qquad (8.53)$$

8.3 Grand partition function of the quantum ideal gases

The integrand vanishes exponentially rapidly with x as $x \to \infty$ because $\exp(-x + \mu/kT)$ becomes very small and $\ln(1+y) \sim y$ for small y. Then, recognizing that $x^{1/2}dx = (2/3)dx^{3/2}$, we may integrate by parts, and get vanishing contributions from the product of the logarithm and $x^{3/2}$ evaluated at the limits of integration, $x = 0$ and $x = \infty$. The result, then, is

$$\ln \Xi = \frac{4g/3}{\sqrt{\pi}\Lambda^3} V \int_0^\infty \frac{x^{3/2} e^{-x+\mu/kT}}{1 \pm e^{-x+\mu/kT}} dx$$

$$= \frac{4g/3}{\sqrt{\pi}\Lambda^3} V \int_0^\infty \frac{x^{3/2}}{e^{x-\mu/kT} \pm 1} dx. \tag{8.54}$$

From (8.54) and (8.12) we then obtain the thermodynamic function $p(T, \mu)$,

$$p(T, \mu) = \frac{4g}{3\sqrt{\pi}} \frac{kT}{\Lambda^3} \int_0^\infty \frac{x^{3/2}}{e^{x-\mu/kT} \pm 1} dx$$

(top sign FD, bottom sign BE). (8.55)

All the intensive properties of the quantum ideal gas then follow from this thermodynamic potential, as, for example, by (8.15)–(8.17).

It is instructive to see what this gives in the Boltzmann statistics, which, as we know, is what would be obtained just by dropping the ± 1:

$$p = \frac{4g}{3\sqrt{\pi}} \frac{kT}{\Lambda^3} e^{\mu/kT} \int_0^\infty x^{3/2} e^{-x} dx$$

$$= g \frac{kT}{\Lambda^3} e^{\mu/kT} \quad \text{(Boltzmann statistics)}. \tag{8.56}$$

(The integral may be found in standard tables; it is $3\sqrt{\pi}/4$.) But from (8.42), $\exp(\mu/kT) = N/z$ in Boltzmann statistics, with z the single-molecule partition function. In the present model z is entirely z_{trans} except for the additional degeneracy factor g; thus, $z = g z_{\text{trans}}$; and z_{trans}, in turn, from (2.17) and the definition of Λ in (8.52), is V/Λ^3. Therefore $\exp(\mu/kT)$ in Boltzmann statistics is $N\Lambda^3/gV$ for this model, and (8.56) is

$$p = NkT/V \quad \text{(Boltzmann statistics)}, \tag{8.57}$$

the equation of state of a perfect gas. But this is not the equation of state of an ideal gas in Fermi–Dirac or Bose–Einstein statistics, where $p(T, \mu)$ is given by (8.55) rather than by (8.56); and where, as we shall learn, the density $\rho(= N/V)$, obtained from $p(T, \mu)$ by (8.15), is not simply $(g/\Lambda^3)\exp(\mu/kT)$.

We now return to the question of the criterion for validity of the Boltzmann statistics, stating it this time in terms of the number of single-particle states with energies within kT of the ground-state energy $\varepsilon_0 \sim 0$, as compared with the number N of molecules. In (8.48) (now to be augmented with the degeneracy factor g) we have the number of these states with energies in the range ε to $\varepsilon + d\varepsilon$, so the number, call it $\Omega(kT)$, with energies between $\varepsilon = \varepsilon_0 \sim 0$ and $\varepsilon = kT$ is

$$\Omega(kT) = \int_0^{kT} \omega(\varepsilon) d\varepsilon$$

$$= 2\pi g V (2m/h^2)^{3/2} \int_0^{kT} \varepsilon^{1/2} d\varepsilon$$

$$= \frac{4\pi}{3} g V \left(\frac{2mkT}{h^2}\right)^{3/2}$$

$$= \frac{4}{3\sqrt{\pi}} g \frac{V}{\Lambda^3} \qquad (8.58)$$

with Λ as given in (8.52). The condition (8.27) for the validity of the Boltzmann statistics, in the thermodynamic limit where $\varepsilon_0 \sim 0$, is $\exp(\mu/kT) \ll 1$; or, since $\exp(\mu/kT) = N\Lambda^3/gV$ in the Boltzmann statistics, as we have just seen, the condition is $N\Lambda^3/gV \ll 1$. Therefore the condition is $N \ll \Omega(kT)$, by (8.58) (since $3\sqrt{\pi}/4$ is of order unity) – the number of particles must be much less than the number of states within kT of the ground state – just as we had supposed.

The condition $N \ll \Omega(kT)$ for the validity of the Boltzmann statistics has itself, in turn, another illuminating interpretation. By (8.58) the condition is

$$\rho \ll 1/\Lambda^3 \quad \text{(for Boltzmann statistics to be valid).} \qquad (8.59)$$

But at number density ρ the volume per molecule is $1/\rho$, so the mean distance between molecules is $\rho^{-1/3}$. Then (8.59) says that the Boltzmann statistics will hold when the mean distance between molecules is much greater than the thermal de Broglie wavelength Λ. That is because with such short wavelengths quantum mechanical diffraction, and so the quantum nature of the particles, becomes unimportant. Another way of looking at it is that the wave function for a pair of particles then has so many nodes between the two particles that it no longer matters whether that number be even or odd; i.e., whether the wave function be even or odd with respect to particle exchange, so whether the particles be fermions or bosons.

The condition (8.59) for the Boltzmann statistics to be a good approximation is certain to hold when, for given temperature (and so for given Λ), the density ρ is small enough, and is certain to fail when the density becomes so large as to be comparable with or greater than $1/\Lambda^3$. Likewise, for given ρ, the condition holds at high temperature T but ultimately fails when the temperature is low enough. How high is "high" and how low is "low" will be seen later.

In the following sections we look separately at the Fermi–Dirac and Bose–Einstein ideal gases, by taking, in turn, the two different signs in (8.55).

8.4 The ideal gas in Fermi–Dirac statistics

Having in mind application to a "gas" of electrons, where $g = 2$, let us now set $g = 2$ and take (8.55) (with the top sign) to be

$$p = \frac{8}{3\sqrt{\pi}} \frac{kT}{\Lambda^3} \int_0^\infty \frac{x^{3/2}}{e^{x-\mu/kT} + 1} \, dx. \tag{8.60}$$

An alternative formula for p, following from $pV = kT \ln \Xi$ [(8.12)] and the formula (8.53) for $\ln \Xi$ before we integrated it by parts to reach (8.54), is

$$p = \frac{4}{\sqrt{\pi}} \frac{kT}{\Lambda^3} \int_0^\infty x^{1/2} \ln\left[1 + e^{-x+\mu/kT}\right] dx. \tag{8.61}$$

The formulas (8.60) and (8.61) for $p(T, \mu)$ are equivalent; the first is the more convenient for calculating p itself but the second is the more convenient for obtaining the number density ρ by differentiation of p, via (8.15). From (8.15) and (8.61), and remembering that Λ, defined by (8.52), depends only on T,

$$\rho = \frac{4}{\sqrt{\pi}} \frac{1}{\Lambda^3} \int_0^\infty \frac{x^{1/2} e^{-x+\mu/kT}}{1 + e^{-x+\mu/kT}} \, dx$$

$$= \frac{4}{\sqrt{\pi}} \frac{1}{\Lambda^3} \int_0^\infty \frac{x^{1/2}}{e^{x-\mu/kT} + 1} \, dx. \tag{8.62}$$

This result may be understood from the formula (8.40) for the mean number of particles occupying a state of energy ε together with the expression (8.48) (with an additional factor $g = 2$) for the number of such states in the energy range ε to $\varepsilon + d\varepsilon$; for they imply that the total number of particles, N, must be

$$N = 4\pi V \left(\frac{2m}{h^2}\right)^{3/2} \int_0^\infty \frac{\varepsilon^{1/2}}{e^{(\varepsilon-\mu)/kT} + 1} \, d\varepsilon. \tag{8.63}$$

Changing the integration variable to $x = \varepsilon/kT$, and with $\rho = N/V$ and Λ as defined in (8.52), we may verify that (8.63) is indeed the same as (8.62).

The energy density u may be obtained from $p(T, \mu)$ via either of the two equations (8.17). There is still a third way to obtain u from p, which is often even more direct. If we think of p/T as being a function of the independent variables $1/T$ and μ, and then recall (8.15) and (8.17), we have

$$\mathrm{d}\frac{p}{T} = \left(\frac{\partial p/T}{\partial \mu}\right)_{1/T} \mathrm{d}\mu + \left(\frac{\partial p/T}{\partial 1/T}\right)_{\mu} \mathrm{d}\frac{1}{T}$$

$$= \frac{1}{T}\rho\,\mathrm{d}\mu + (\rho\mu - u)\mathrm{d}\frac{1}{T}$$

$$= \rho\,\mathrm{d}\frac{\mu}{T} - u\,\mathrm{d}\frac{1}{T}, \qquad (8.64)$$

because $\mathrm{d}(\mu/T) = (1/T)\,\mathrm{d}\mu + \mu\,\mathrm{d}(1/T)$. Therefore, now thinking of p/T as a function of $1/T$ and μ/T,

$$u = -\left(\frac{\partial p/T}{\partial 1/T}\right)_{\mu/T}; \qquad (8.65)$$

i.e., we may obtain u just by differentiating p/T with respect to $1/T$ at fixed μ/T.

This is particularly convenient here because the integral in (8.60) is a function only of μ/T; at fixed μ/T the dependence of p/T on $1/T$ comes entirely from the factor $1/\Lambda^3$. With Λ given by (8.52), $1/\Lambda^3$ is seen to be proportional to $T^{3/2}$; i.e., to $(1/T)^{-3/2}$. Differentiating this with respect to $1/T$ gives $-(3/2)(1/T)^{-5/2} = -(3/2)T(1/T)^{-3/2}$; i.e., the effect of differentiating $1/\Lambda^3$ with respect to $1/T$ is merely to multiply it by $-(3/2)T$. Therefore, from (8.60), the derivative of p/T with respect to $1/T$ at fixed μ/T is just $(-3/2)T(p/T) = -(3/2)p$. Then from (8.65) the energy density u is just

$$u = \frac{3}{2}p. \qquad (8.66)$$

This is a remarkable result. In the Bose–Einstein and Boltzmann statistics, too, according to (8.55) and (8.56), p/T at fixed μ/T depends on T only through the factor $1/\Lambda^3$, so the same result (8.66) holds there, too; i.e., the relation $u = (3/2)p$ between the pressure and the energy density holds for any ideal gas of particles with no excitable internal degrees of freedom, whether in the Fermi–Dirac, the Bose–Einstein, or the Boltzmann statistics. The energy density u in these circumstances is the density of translational kinetic energy alone. This $u = (3/2)p$ is just the relation $pV = (2/3) \times$ (total kinetic energy of translation) familiar from the kinetic theory of gases. What is remarkable is that it continues to hold in the Fermi–Dirac and Bose–Einstein statistics even though the two separate relations $p = \rho k T$ and $u = (3/2)\rho k T$ hold only in Boltzmann statistics.

8.4 The ideal gas in Fermi–Dirac statistics

Just as the formula (8.62) for ρ in the Fermi–Dirac statistics could be understood as a reflection of the expressions (8.40) and (8.48) or (8.49) for the occupation numbers and the density of states, so also can the formula (8.66) for u be so understood when it is combined with (8.60) for p. Thus, it ought to be so that

$$u = \frac{1}{V} \int_0^\infty \varepsilon \bar{n}_\varepsilon \omega(\varepsilon) d\varepsilon; \qquad (8.67)$$

and inserting (8.40) here for $\bar{n}_\varepsilon$ (with the former ε_r now understood to be the integration variable ε) and (8.49) (but now augmented by the extra factor $g = 2$) for $\omega(\varepsilon)$, and recalling the definition of Λ in (8.52) and the expression for p in (8.60), we may indeed see that (8.66) for u is the same as (8.67).

We have remarked that it is not the case that $p = \rho kT$ or $u = (3/2)\rho kT$ in Fermi–Dirac statistics; instead, while it is still true that $u = (3/2)p$, p and ρ are now given as functions of T and μ by (8.60) and (8.62). For general μ/T the integrals in those formulas cannot be simplified, but it is instructive to look at the extremes of large positive and large negative values of μ/T.

The latter extreme, $\mu/kT \ll -1$, is that in which the density is very low relative to $1/\Lambda^3$ [recall $\rho = (g/\Lambda^3) \exp(\mu/kT)$ in Boltzmann statistics] and so is the limit in which the Boltzmann statistics should hold. Indeed, as we see in (8.60) and (8.62), the $+1$ in the denominators of the integrands become negligible compared with $x \exp(-\mu/kT)$ as $\mu/kT \to -\infty$, and we obtain

$$p \sim \frac{8}{3\sqrt{\pi}} \frac{kT}{\Lambda^3} e^{\mu/kT} \int_0^\infty x^{3/2} e^{-x} dx$$

$$= \frac{2kT}{\Lambda^3} e^{\mu/kT} \quad (\mu/kT \to -\infty) \qquad (8.68)$$

[cf. (8.56) with $g = 2$] and

$$\rho \sim \frac{4}{\sqrt{\pi}} \frac{1}{\Lambda^3} e^{\mu/kT} \int_0^\infty x^{1/2} e^{-x} dx$$

$$= \frac{2}{\Lambda^3} e^{\mu/kT} \quad (\mu/kT \to -\infty) \qquad (8.69)$$

[cf. the earlier $\rho = (g/\Lambda^3) \exp(\mu/kT)$ in the Boltzmann statistics, with $g = 2$]. Together, these give the expected $p = \rho kT$.

Much more interesting is the opposite extreme, $\mu/kT \gg 1$, which is also that of high density or low temperature. Let $y = (kT/\mu)x$ be a new variable of integration in (8.60), so that

$$p = \frac{8}{3\sqrt{\pi}} \frac{kT}{\Lambda^3} \left(\frac{\mu}{kT}\right)^{5/2} \int_0^\infty \frac{y^{3/2}}{e^{(\mu/kT)(y-1)} + 1} dy. \tag{8.70}$$

Then as $\mu/kT \to \infty$ the integrand becomes just $y^{3/2}$ for $y < 1$ and 0 for $y > 1$. Then

$$p \sim \frac{8}{3\sqrt{\pi}} \frac{kT}{\Lambda^3} \left(\frac{\mu}{kT}\right)^{5/2} \int_0^1 y^{3/2} dy$$

$$= \frac{16}{15\sqrt{\pi}} \frac{kT}{\Lambda^3} \left(\frac{\mu}{kT}\right)^{5/2}$$

$$= \frac{16\pi}{15} \left(\frac{2m}{h^2}\right)^{3/2} \mu^{5/2} \quad (\mu/kT \to \infty) \tag{8.71}$$

where we have made use of the definition of Λ in (8.52). Note that in this limit p is independent of T at given μ. Similarly, from (8.62),

$$\rho \sim \frac{4}{\sqrt{\pi}} \frac{1}{\Lambda^3} \left(\frac{\mu}{kT}\right)^{3/2} \int_0^1 y^{1/2} dy$$

$$= \frac{8}{3\sqrt{\pi}} \frac{1}{\Lambda^3} \left(\frac{\mu}{kT}\right)^{3/2}$$

$$= \frac{8\pi}{3} \left(\frac{2m}{h^2}\right)^{3/2} \mu^{3/2} \quad (\mu/kT \to \infty), \tag{8.72}$$

again independent of T at given μ. In this extreme, then, we have for the Fermi–Dirac ideal gas

$$p \sim \frac{16\pi}{15} \left(\frac{2m}{h^2}\right)^{3/2} \left[\frac{3}{8\pi} \left(\frac{h^2}{2m}\right)^{3/2} \rho\right]^{5/3}$$

$$= \frac{1}{20} \left(\frac{3}{\pi}\right)^{2/3} \frac{h^2}{m} \rho^{5/3} \quad (\mu/kT \to \infty). \tag{8.73}$$

This is what replaces $p = \rho kT$ of the Boltzmann statistics in the extreme of very high density or very low temperature. Note now the proportionality of p to the 5/3 power of ρ instead of to the first power, and note that in this limit the pressure at fixed ρ is independent of T instead of proportional to it. From (8.66), the energy density in this same limit is

$$u \sim \frac{3}{40} \left(\frac{3}{\pi}\right)^{2/3} \frac{h^2}{m} \rho^{5/3} \quad (\mu/kT \to \infty). \tag{8.74}$$

8.4 The ideal gas in Fermi–Dirac statistics

To understand how these have come about we should ask what happens to the mean occupation number $\bar{n}_\varepsilon$ in this same high-density, low-temperature limit, which is the limit $\mu/kT \to \infty$. What we here (and earlier, in (8.67)) call $\bar{n}_\varepsilon$ is the same as the $\bar{n}_r$ of (8.40), now with ε_r simply called ε; so $\bar{n}_\varepsilon$ is the mean number of particles (electrons, say, with $g = 2$) in a single-particle state of energy ε. Then from (8.40), in the Fermi–Dirac statistics,

$$\bar{n}_\varepsilon = \frac{1}{e^{(\mu/kT)(\varepsilon/\mu - 1)} + 1}; \tag{8.75}$$

so for $\mu/kT \gg 1$,

$$\bar{n}_\varepsilon \sim \begin{cases} 1, & \varepsilon < \mu \\ 0, & \varepsilon > \mu \end{cases} \quad (\mu/kT \gg 1). \tag{8.76}$$

This is essentially the observation we already made when we approximated the integrand in the formula (8.70) for p by that in (8.71). The occupation number $\bar{n}_\varepsilon$ is shown as a function of ε in Fig. 8.4. The solid lines are from (8.76) and represent the $T = 0$ extreme for any given μ (so that $\mu/kT = \infty$), while the dashed curve, on which the corners are rounded, corresponds to some $T > 0$ at the same μ. At $T = 0$, every single-particle state of energy $\varepsilon < \mu$ contains one particle and every state of energy $\varepsilon > \mu$ contains none.

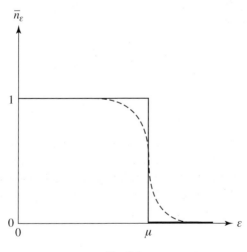

Fig. 8.4

At $T = 0$ the system as a whole must be in its ground state. For fermions, for which multiple occupancy of the single-particle states is forbidden, the ground state of the system is that in which one particle is in the state $r = 0$, one in the state $r = 1$, etc., until each of the N lowest-lying states contains one particle, while no state of higher energy contains any. This is exactly what Fig. 8.4 says, provided that at $T = 0$ the total number of single-particle states of energy $\leq \mu$ is the same as the total number of particles, N. To check the consistency of this interpretation, then, we should calculate this number of single-particle states of energy $\leq \mu$; call it $\Omega(\mu)$. From the density of states $\omega(\varepsilon)$ given in (8.49), but now augmented by the extra factor $g = 2$,

$$\Omega(\mu) = \int_0^\mu \omega(\varepsilon) d\varepsilon = \frac{8\pi}{3} V \left(\frac{2m}{h^2}\right)^{3/2} \mu^{3/2}. \tag{8.77}$$

Comparing this with the limiting density ρ in (8.72) we see that in this limit $\Omega(\mu) = \rho V$, which is indeed N.

The level ε up to which, at low temperatures, each state is singly occupied and beyond which each is empty, is called the *Fermi level*, ε_F. We see from Fig. 8.4 that the chemical potential of the system at $T = 0$ is the Fermi energy ε_F.

When the temperature is low enough for $\bar{n}_\varepsilon$ to be accurately approximated by the step function in (8.76) or Fig. 8.4 the gas is said to be in its fully *degenerate* state. That is when the Fermi–Dirac statistics is most strongly manifest.

How close are we to that condition under normal circumstances – say, for an electron "gas" at the density of the valence electrons in a metal, and at room temperature? It is (8.59) that provides the criterion: when the density is so high or the temperature so low that $\rho \gg 1/\Lambda^3$ the gas is highly degenerate, while if $\rho \ll 1/\Lambda^3$ we are at the opposite extreme, where the Fermi–Dirac statistics is unimportant. The borderline case is $\rho \simeq 1/\Lambda^3$. The density of the valence electrons in a metal is of order 1 Å^{-3}. With Λ given by (8.52), we may calculate from the known mass of the electron that the temperature at which $\rho \Lambda^3 = 1$ when $\rho = 1$ Å^{-3} is about 500,000 K [Exercise (8.6), below]. At any temperature much lower than that – hence, at any temperature of practical interest – the electron gas is almost fully degenerate. That the chemical potential of the valence electrons is then identifiable with the Fermi energy ε_F is an important fact of electrochemistry.

Exercise (8.4). Calculate the mean energy per electron, $\bar{\varepsilon}$, in the fully degenerate electron gas, expressing it in terms of the Fermi energy ε_F.

8.4 The ideal gas in Fermi–Dirac statistics

Solution. The mean energy per electron is the ratio u/ρ of the energy density to the number density. Since $u = (3/2)p$ this is also $(3/2)p/\rho$, so from (8.71) and (8.72),

$$\bar{\varepsilon} = \frac{3}{5}\mu = \frac{3}{5}\varepsilon_F.$$

Note that this is more than $(1/2)\varepsilon_F$ but less than ε_F. As seen in (8.48), the density of states $\omega(\varepsilon)$ increases with increasing ε (proportionally to $\sqrt{\varepsilon}$), so there are more states of higher energy than of lower. Therefore more than half the particles in the fully degenerate gas have energy greater than $(1/2)\varepsilon_F$. Hence, $\bar{\varepsilon}$ has to be greater than $(1/2)\varepsilon_F$, but it must be less than the maximum energy, ε_F, of the occupied levels. Since in the degenerate limit each state of energy $\varepsilon \leq \varepsilon_F$ contains one electron and no state of energy $\varepsilon > \varepsilon_F$ contains any, the result $\bar{\varepsilon} = (3/5)\varepsilon_F$ may also be obtained directly as

$$\bar{\varepsilon} = \int_0^{\varepsilon_F} \varepsilon\omega(\varepsilon)d\varepsilon \bigg/ \int_0^{\varepsilon_F} \omega(\varepsilon)d\varepsilon$$

$$= \int_0^{\varepsilon_F} \varepsilon^{3/2}d\varepsilon \bigg/ \int_0^{\varepsilon_F} \varepsilon^{1/2}d\varepsilon$$

$$= \frac{2}{5}\varepsilon_F^{5/2} \bigg/ \frac{2}{3}\varepsilon_F^{3/2} = \frac{3}{5}\varepsilon_F.$$

Exercise (8.5). Show that the product Ts of the absolute temperature and the entropy density is 0 in the fully degenerate electron gas.

Solution. From the first of Eqs. (8.17),

$$Ts = u - \rho\mu + p.$$

Since $u = (3/2)p$ this is

$$Ts = (5/2)p - \rho\mu,$$

so from (8.71) and (8.72)

$$Ts = 0.$$

To see that it is not only Ts that is 0 in the fully degenerate limit but also s itself, in agreement with the third law (Chapter 5), requires going beyond just the first approximation (8.71) to the general formula (8.60) or (8.70) for p.

Exercise (8.6). With Λ the thermal de Broglie wavelength of electrons, calculate the temperature at which $\rho\Lambda^3 = 1$ when $\rho = 1$ Å^{-3}.

Solution. This is the temperature at which $\Lambda = 1$ Å:

$$\frac{h}{\sqrt{2\pi m_e kT}} = 10^{-8} \text{ cm} = 10^{-10} \text{ m}$$

with

$$m_e = 9.109 \times 10^{-31} \text{ kg}.$$

Then

$$T = \frac{h^2}{2\pi m_e k (10^{-10} \text{ m})^2}$$

$$= \frac{(6.626 \times 10^{-34} \text{ J s})^2}{2\pi (9.109 \times 10^{-31} \text{ kg})(1.3807 \times 10^{-23} \text{ J/K})(10^{-10} \text{ m})^2}$$

$$= 556,000 \text{ K}.$$

(This is the temperature that is said in the text to be "about 500,000 K.")

8.5 The ideal gas in Bose–Einstein statistics

We return to the formulas (8.40) and (8.55) for the occupation numbers $\bar{n}_r$ and the pressure $p(T, \mu)$, but now we take the bottom sign.

Since a negative $\bar{n}_r$ is an impossibility, we conclude immediately from (8.40) that μ can never be greater than the smallest ε_r, which is the ground-state energy ε_0. But ε_0 is itself 0 in the thermodynamic limit, with the zero of energy we have adopted; so we see that, referred to that scale, the chemical potential of the ideal Bose gas can never be positive. This is wholly different from the Fermi gas and the Boltzmann gas, in both of which μ can be of either sign. This physical limit on the possible values of μ gives to the Bose gas unique properties, as we shall see.

Having in mind eventual application to ^{4}He, we take the degeneracy factor g to be 1. Then from (8.55), the thermodynamic potential $p(T, \mu)$ of the ideal Bose gas is

$$p = \frac{4}{3\sqrt{\pi}} \frac{kT}{\Lambda^3} \int_0^\infty \frac{x^{3/2}}{e^{x-\mu/kT} - 1} dx. \tag{8.78}$$

As remarked in §8.4, the relation (8.66) between the energy density u and the pressure p holds equally well in the Bose–Einstein, the Fermi–Dirac, and the

8.5 The ideal gas in Bose–Einstein statistics

Boltzmann statistics, so we have here, too,

$$u = \frac{3}{2}p. \tag{8.79}$$

Just as the resulting expression for u in the Fermi–Dirac case could be understood from (8.67), with $\bar{n}_\varepsilon$ from (8.40) and $\omega(\varepsilon)$ from (8.49) (with an extra factor $g = 2$), so also here the expression for u that results from $u = (3/2)p$ with p as in (8.78) can be understood from the same (8.67), now taking the bottom sign in (8.40) for $\bar{n}_\varepsilon$.

If the density ρ were calculated from this $p(T, \mu)$ by means of (8.15) we would simply obtain the analog of (8.62), which is the formula for ρ in the Fermi–Dirac statistics, but now with -1 instead of $+1$ in the denominator of the integrand, and with $g = 1$ instead of $g = 2$ so that the factor preceding the integral would be $2/(\sqrt{\pi}\Lambda^3)$. Let us record that formula here, but with an additional term called ρ_0 on the right-hand side, for reasons that will be apparent shortly; thus,

$$\rho = \frac{2}{\sqrt{\pi}} \frac{1}{\Lambda^3} \int_0^\infty \frac{x^{1/2}}{e^{x-\mu/kT} - 1} dx + \rho_0. \tag{8.80}$$

Once again, as in the Fermi–Dirac gas, this formula for ρ *except for* the added term ρ_0 can be understood simply as arising from

$$N = \int_0^\infty \bar{n}_\varepsilon \omega(\varepsilon) d\varepsilon \tag{8.81}$$

with (8.40) for $\bar{n}_\varepsilon$ and (8.49) for $\omega(\varepsilon)$ [cf. (8.63) in the Fermi–Dirac case].

Why is the first term on the right-hand side of (8.80) then not the whole of ρ? When and why is there an additional ρ_0? The answer comes from considering what happens to the integral in (8.80) as μ increases through negative values to its maximum physically allowed value, $\mu = 0$. This is shown in Fig. 8.5, which

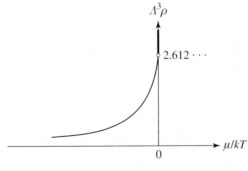

Fig. 8.5

for $\mu/kT < 0$ is a schematic plot of $\Lambda^3\rho$ as a function of μ/kT as given by (8.80) with no extra term ρ_0. The limiting value of $\Lambda^3\rho$ as μ/kT approaches 0 from negative values is shown as $2.612\cdots$, which is the value of

$$\frac{2}{\sqrt{\pi}}\int_0^\infty \frac{x^{1/2}}{e^x - 1}dx = 2.612\cdots. \tag{8.82}$$

[If the power of x in the integrand were more generally some power $s - 1$, the integral would be expressible in terms of what in the mathematical literature is called the "zeta function," $\zeta(s)$; the left-hand side of (8.82) is $\zeta(3/2) = 2.612\cdots$.] But there is no physical limit to how great ρ may be; this is a gas of non-interacting particles, and an arbitrarily large number of them can be accommodated in any volume. What, then, is the rest of the curve of $\Lambda^3\rho$ versus μ/kT in Fig. 8.5 like, when $\Lambda^3\rho > 2.612\cdots$? We know that μ/kT cannot become positive. Nor can the curve double back to negative values of μ/kT as $\Lambda^3\rho$ increases beyond $2.612\cdots$ because $(\partial\mu/\partial\rho)_T$ can never be negative in any stable thermodynamic equilibrium state. [This is equivalent to the condition that the isothermal compressibility, which is $\rho^{-1}(\partial\rho/\partial p)_T = \rho^{-2}(\partial\rho/\partial\mu)_T$, can never be negative.] Therefore the curve can do nothing but stick at $\mu/kT = 0$, so it just continues up the vertical axis as $\Lambda^3\rho$ increases beyond $2.612\cdots$. This is indicated in Fig. 8.5 by a thickened line.

In the meantime, since μ/kT is stuck at 0, the first term on the right-hand side of (8.80) is stuck at $2.612\cdots/\Lambda^3$ even while ρ itself increases beyond that. It is that additional density, which is not accounted for by the integral in (8.80) and which does not appear until $\Lambda^3\rho$ exceeds $2.612\cdots$, that we are calling ρ_0. Thus,

$$\rho_0 = \begin{cases} 0, & \rho \leq (2.612\cdots)/\Lambda^3 \\ \rho - (2.612\cdots)/\Lambda^3, & \rho \geq (2.612\cdots)/\Lambda^3. \end{cases} \tag{8.83}$$

Figure 8.6 shows ρ_0/ρ, the fraction of the total density that is contributed by ρ_0,

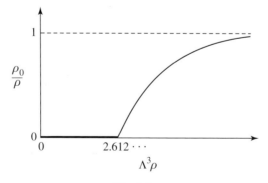

Fig. 8.6

8.5 The ideal gas in Bose–Einstein statistics

as a function of $\Lambda^3\rho$, from (8.83). Until $\Lambda^3\rho$ reaches $2.612\cdots$ there is no ρ_0, but then a ρ_0 appears and grows. As $\Lambda^3\rho$ continues to increase, this ρ_0 contributes an increasingly large fraction of the total ρ, approaching the whole of it as $\Lambda^3\rho$ becomes very great.

Why did the integral in the first term in (8.80) miss this extra density ρ_0, thus requiring it to be added separately? The reason is that while $\bar{n}_\varepsilon \omega(\varepsilon)d\varepsilon$ with $\bar{n}_\varepsilon$ given by (8.40) and $\omega(\varepsilon)$ by (8.49) gives correctly the infinitesimal number of particles with energies in the range ε to $\varepsilon + d\varepsilon$ for any $\varepsilon > 0$ (cf. (8.81)), it misses a *macroscopic* occupancy of the ground state – the $\varepsilon = 0$ state, in the thermodynamic limit – that begins when $\Lambda^3\rho$ reaches $2.612\cdots$. Until that point is reached the number of particles with energies within $d\varepsilon$ of $\varepsilon = 0$ is infinitesimal, but once $\Lambda^3\rho$ reaches that critical value, i.e., once μ becomes 0, any further increase in density or decrease in temperature causes a macroscopic number of particles, $(\rho_0/\rho)N$, which is $O(N)$, to occupy the states that in the thermodynamic limit ($V \to \infty$) are of energy $\varepsilon = 0$.

Since the particles that occupy the $\varepsilon = 0$ state have vanishing energy and momentum they do not contribute to either the energy density u or the pressure p. Therefore the formulas (8.78) and (8.79) for p and u remain correct, with no added terms, even after $\Lambda^3\rho$ increases beyond $2.612\cdots$. At larger $\Lambda^3\rho$ both $\Lambda^3 p/kT$ and $\Lambda^3 u/kT$ stick at the values they had when $\Lambda^3\rho$ reached that critical value, just as μ sticks at 0 (Fig. 8.5). This is shown for $\Lambda^3 p/kT$ in Fig. 8.7. (The number $1.341\cdots$ is $\zeta(5/2)$.) With an extra factor of $3/2$ this would be a plot of $\Lambda^3 u/kT$. The dashed line of slope 1 is the ideal-gas law, which holds when $\Lambda^3\rho \ll 1$. We see that in the ideal Bose gas's *degenerate* state ($\mu = 0$, $\Lambda^3\rho > 2.612\cdots$) the pressure and energy density depend only on

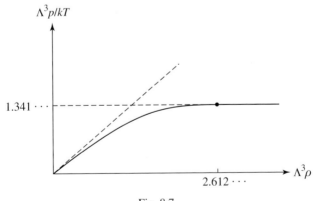

Fig. 8.7

the temperature; for given temperature they are independent of the density. This is the exact opposite of what we found for the ideal Fermi gas in *its* degenerate state, in (8.73) and (8.74), where p and u proved to be proportional to $\rho^{5/3}$ and independent of T.

This behavior of the ideal Bose gas was first inferred by Einstein, and was then shown even more convincingly by F. London (the same London after whom the attractive component of van der Waals forces is named; §6.2). It is referred to as Einstein *condensation*: an accumulation of particles in a state in which they have no momentum. London noted that something closely analogous to Einstein condensation should occur in liquid ^{4}He, just as a consequence of the Bose–Einstein statistics; and that this could account for the transition to a "superfluid" state, and the attendant thermodynamic singularities, that occurs when liquid helium, under the pressure of its own vapor, is cooled below 2.17 K. In that state the liquid flows without viscous resistance (hence the name "superfluid") if the flow velocity is not too great. (It is thought that the transition to superconductivity – the flow of electricity without resistance – in metals at very low temperatures has its origin in an analogous Bose–Einstein condensation of electron *pairs*.) Near the transition to that state a plot of the heat capacity of the liquid as a function of temperature has a shape somewhat reminiscent of the Greek letter lambda, so that point is often called the "lambda point" of helium. Figure 8.8 shows such a plot of the molar constant-volume heat capacity C_V of

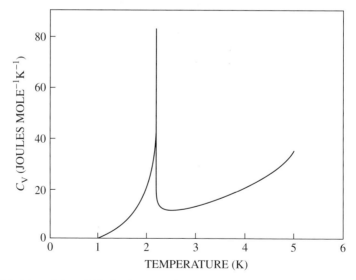

Fig. 8.8 Adapted from M.R. Moldover and W.A. Little, in '*Critical Phenomena*,' National Bureau of Standards Misc. Publ. **273** (1966), M.S. Green and J.V. Sengers, eds., pp. 79–82. Reproduced with permission.

8.5 The ideal gas in Bose–Einstein statistics

liquid helium as a function of temperature, with the liquid under the pressure of its vapor.

Liquid helium is far from being an ideal gas. The singularities in its thermodynamic functions at the lambda point are strongly affected by the interatomic interactions and so are not the same as in the ideal gas – seen here in Figs. 8.5–8.7, for example. Thus, the C_V vs. T curve for the ideal gas, while showing an obvious singularity at the temperature T at which $\Lambda^3 \rho = 2.612 \cdots$ for any given ρ, looks rather different from the plot for real liquid helium in Fig. 8.8. Nevertheless, it is striking how close the temperature of the transition in the ideal gas is to the temperature of the superfluid transition in liquid helium: an ideal Bose gas of particles of mass equal to that of ^{4}He atoms, at the density of liquid helium, undergoes its transition at 3.15 K. This is close enough to the lambda point of liquid helium, 2.17 K, for it to be believable that the superfluid transition is due to Einstein condensation and is thus a manifestation of the Bose–Einstein statistics.

Index

argon, 83, 84, 102–4, 106, 111, 112

barometric distribution, vii, 1, 3
black-body radiation, 55, 66
Boltzmann constant, 2, 4, 14
Boltzmann distribution law, vii, viii, 1–4, 7, 10, 19, 42, 88, 127, 130, 131, 138, 147
Boltzmann statistics, 16, 19, 20, 49, 133, 134, 136, 145–7, 149, 150, 153, 154, 156–8, 162, 163
Born–Oppenheimer approximation, 27
Bose–Einstein condensation, 166
Bose–Einstein statistics, 20, 21, 133, 134, 145–50, 152, 153, 155, 156, 162, 166, 167
boson, 20, 34, 133–6, 143, 145, 147, 148
Boyle temperature, 82

Carnahan–Starling equation, 110–12
colloid, 4, 86
computer simulation, 103, 104, 110, 111, 114–18, 127
computing, 102, 114, 115, 127, 130, 132
configuration integral, 128
configuration space, 129–31
continuum, elastic, 57, 62, 66, 67, 151
critical point, 112

Debye theory, 62, 64, 76
Debye–Hückel theory, 100
degenerate quantum gas, 160, 161, 165
delta function, 125
density of states, 9, 10, 13, 15, 19, 71–4, 79, 151, 152, 157, 161, 163, 165
distribution function
  energy-probability, 10–12, 139–41
  number-probability, 139–41, 147, 148
  pair (radial), 88–90, 102–6, 110, 113, 117, 118, 124, 126, 132
Dulong–Petit law, 31, 63, 64, 66

Einstein condensation, 166, 167
Einstein model, 66, 73
electron gas (electrons in metals), viii, 20, 21, 133, 160
electron spin, 151, 152
electrons, valence, 20, 133, 160
energy density, 142, 143, 155–8, 161–3, 165, 166
entropy density, 142, 143, 161
entropy of mixing, 26
entropy, residual, 79, 80
equilibrium constant, viii, 47, 49, 51, 53, 131
equipartition law, 41, 42, 63, 66, 68, 119

Fermi–Dirac statistics, 20, 21, 133, 134, 145–50, 152, 153, 155–60, 162, 163, 166
Fermi energy (level), 160
fermion, 20, 34, 133–6, 143, 145, 147, 148, 160
field variable, 139
finite-size effects, 115, 116, 127, 130
fluctuations, viii, 6, 7, 10, 11, 13, 140, 141
force, intermolecular
  attractive, 84–9, 94, 95, 98, 102, 104, 106, 166
  Coulomb, 98, 99
  dispersion, 86, 98
  London, 86, 98
  long-ranged, 98
  repulsive, 84, 85, 87–9, 94, 95, 102, 104, 106
  retardation, 86
  short-ranged, 98
frequency distribution, 55–7, 61–7, 151

gas, non-ideal (imperfect), viii, 2, 76, 79, 81, 82, 88, 96–8, 100, 112, 120
Gaussian distribution, 12
Gibbs–Duhem equation, 142
Gibbs–Helmholtz equation, viii, 7, 8, 24, 28

gravity, 3, 4
Green's theorem, 121

hard spheres, 84, 102–9, 111, 113–16, 124–6, 130, 132
heat capacity, 24, 30, 31, 36–8, 40, 43, 45, 62–6, 75, 166, 167
helium, 20, 21, 70, 101, 133, 134, 150, 162, 166, 167
hydrogen
    bond, 80
    ortho- and *para-*, vii, 31, 35–8

importance sampling, 129, 130
ionic activity coefficient, 100
ionic strength, 100
ionosphere, 99
isochore, 108

Joule–Thomson inversion temperature, 82

kinetic theory, 4, 118, 152, 156

lambda point, 166, 167
Laplace transform, 9, 15
Lennard-Jones potential, 86, 96
liquids, viii, 2, 88, 90, 101, 102, 106–8, 112, 113, 117, 120, 129, 132
London theory, 166

magnetosphere, 100
mass-action law, 49
Maxwell velocity distribution, vii, 1–3, 55, 117, 120
Metropolis algorithm, 130, 131
molecular dynamics, vii, 114–16, 124, 130–2
Monte Carlo, vii, 127, 131

Nernst–Einstein relation, viii
Nernst heat theorem, 69, 71
neutron scattering, 102
normal modes, 27, 58, 59, 61, 62, 101, 151
nuclear fusion, 99
nuclear spin, 34–7, 74

occupation numbers, 134, 135, 143, 147–50, 155, 157, 159, 160, 162, 163, 165
osmotic pressure, 100

particle-in-a-box, 9, 19, 22, 151
partition function
    canonical, viii, 15, 138, 141, 143–5
    electronic, 43, 44, 52
    grand, vii, 138, 139, 141–7, 150, 152

internal, 21, 23, 31, 43–6, 50, 128
microcanonical, viii, 15, 138, 141
rotational, 27, 31–40, 44, 45, 52, 76, 77
single-molecule, 17, 18, 21, 23, 27, 49, 50, 133, 136, 146, 149, 150, 153
translational, 21–4, 27, 32, 45, 50, 52, 128, 153
vibrational, 27–30, 44, 45, 52, 53, 62, 67, 77
Pauli exclusion principle, 20, 84
periodic boundary conditions, 116, 130
Planck radiation law, 66–8

quantum number, 5, 16, 18, 20–22

random number, 130, 131
random walk, 11
Rayleigh–Jeans law, 57, 61, 63, 66, 151
Rayleigh radiation law, 68

Sackur–Tetrode equation, 26, 76, 77
solid, harmonic, viii, 55, 101
square-well potential, 87, 93, 96, 113, 114, 124
stellar interior, 99
Stirling approximation, 25
superconductivity, 166
superfluid, 166, 167
symmetry number, 33–5, 38, 39, 76

thermal expansion, 63
thermal (de Broglie) wavelength, 152, 154, 162
thermodynamic limit, 61, 72, 73, 91, 151, 152, 154, 162, 165
thermodynamic potential, 8, 15, 138, 141, 142, 153, 162
thermostat, 6, 140
third law of thermodynamics, viii, 69, 71, 73, 74, 76, 78, 98, 101
triple point, 102, 112, 113

ultraviolet catastrophe, 68

van der Waals equation, 82, 84, 108, 111–13
van der Waals forces, 84, 102, 166
van der Waals parameters, 85, 94, 95, 106, 108, 113
virial coefficient, 81–4, 87, 88, 90, 93–9, 109, 110, 113
virial series, 81, 89, 93, 97, 99
virial theorem, 81, 118, 119, 123, 124

waves, 57–9, 61, 66

zero-point energy, 29, 31, 51, 151, 152
zeta function, 164, 165